# Integrated Design and Cost Management for Civil Engineers

# Integrated Design and Cost Management for Civil Engineers

Editor

**Dilip Warade**

scitus
academics

**Integrated Design and Cost Management for Civil Engineers**

Edited by **Dilip Warade**

Printed in 2017

ISBN: 978-1-68117-011-4

Library of Congress Control Number: 2015931819



SCITUS Academics LLC,
616, Corporate Way, Suite 2, 4766,
Valley Cottage, NY 10989

www.scitusacademics.com

# Contents

# Preface

Milestone cost estimates at various stages of the process are critical components of the cost management activity. Risk Assessment and Management are important as identi ed risks on construction projects are typically nancial in nature. Therefore early in the project an assessment of risk is crucial to establish the budget parameters within which the project must be completed. The calculation of project contingencies should be based on an assessment of the risk surrounding the project (site issues, availability of bidders, method of procurement, and critically the market conditions in the location of the project. As risks are mitigated then contingencies can be reduced and the range of estimated nal cost narrowed. The rm charged with managing the costs of the project should ideally be hired directly by the owner, early in the process, and should be independent of both the architect/ engineer and the construction contractor. Throughout a project's planning, design, and construction phases, Cost Management is employed as a means of balancing a project's scope and expectations of quality and budget. The book text details procedures that emphasize the correct quantication and planning of works, resulting in reliable cost and time predictions. This book is a reference tool to design and cost management in civil engineering. This book is an ideal textbook for undergraduate civil engineering and quantity surveying students, as well as postgraduate civil engineering management students.

**Editor**

Chapter 1

# Causes of Variation Orders in Road Construction Projects in Sri Lanka

R. U. Halwatura and N. P. N. P. Ranasinghe

Department of Civil Engineering, University of Moratuwa, 10400 Moratuwa, Sri Lanka

## ABSTRACT

One of the most important problems in the construction industry is variations. They occur in every construction project and the magnitude of these variations varies considerably from project to project. Hence, the variations orders bear great importance right from the inception to completion in the construction industry. Most of the road construction projects in Sri Lanka have experienced a large number of variation orders. The client had to spend more than what was initially estimated in most cases. Sometimes, disputes and unnecessary delays occur due to variations. This study attempted to reveal the possible causes of variation orders in the road construction projects in Sri Lanka. The data were collected through a literature review, a case study analysis focused on 11 road construction projects, and a questionnaire administered to professionals in the road construction industry in Sri Lanka. The study found out that the causes in the local context differ from those in the international context. According to the questionnaire survey, poor estimation was the most significant cause of variation orders. Unforeseen site conditions, political pressure during construction stage, poor investigation, and client-initiated variations occupy the 2nd to 5th places, respectively, in the ranking. This ranking was further proven through the case study analysis.

## INTRODUCTION

The cost of a construction project is one of the most important factors in the construction industry. Due to many reasons, the total cost of a project can significantly vary from the initial estimated cost. The reasons could be changes in scope of work, specifications, or any other contract documents. In the construction industry, variation orders are created when changes occur. It is an official document that states the changes made into the original agreement between the client and the contractor. When a variation order is created, it brings several negative effects to both the client and the contractor.

The study discussed in this paper was aimed at investigating possible causes of variation orders in the road construction projects in Sri Lanka. The data for the analysis were collected using three methods. The causes of variation orders in general in the road construction industry and their ranking were identified based on a literature review. Then the findings were surveyed based on a questionnaire administered to the professionals in the industry in Sri Lanka and selective case studies were analyzed to understand the degree of occurrence and their impact.

This paper first provides a brief literature review to discuss the theoretical background to the study and then it discusses the methodology used in the study, results with discussion, and finally the main conclusions with some recommendations.

# THE LITERATURE REVIEW

The cost of a construction project right from inception to completion bears great importance in the construction industry. However, due to many reasons, the total cost of a project can significantly vary from the initial estimated cost. These variations are inevitable in any construction project [1]. Fisk [2] defines the following: "a variation is any deviation from an agreed well-defined scope and schedule. Stated differently, this is a change in any modification to the contractual guidance provided to the contractor by the owner or owner's representative." This includes changes to plans, specifications, or any other contract documents.

Clearly, in construction terms, variability is referred to as variation orders. A variation order is the formal document that is used to modify the original contractual agreement provided to the contractor by the client or the client's representative and becomes part of the project's documents [3–5]. Arguably, variation orders may be seen as counter to the principle of waste reduction. The more the variation orders are on a project, the greater the likelihood that they become time consuming and costly elements in construction projects is [6]. When a variation order occurred, the contractor tends to charge higher rates for variation items. Then the client is affected in terms of cost.

Success in managing variation orders results in uninterrupted construction operations and agreed project costs as well as duration [7]. However, this is not always practically achievable. A variation order has to be managed carefully. Otherwise, disputes between a client and a contractor related to cost and time of work might occur [7]. Variation orders often involve additional cost and disruption to work already underway, leading to cost and time overruns, quality degradation, and loss in productivity on construction projects [8].

Previous studies related to variation orders have mainly focused on management aspects and their overall costs [6, 7, 9]. A study of delays and cost increases in the construction of private residential projects in Kuwait revealed that a number of variation orders issued during the construction phase led to both delays and cost increases [10]. The projects that experienced variation orders incurred more than 58% time delay and cost increases when compared to those with no variation orders [10]. The magnitude of variation orders varies from one project to another. Though there have been cases where variation costs accounted for as much as 100% of the budgeted funds, the industry norm has been determined to be about 10% [5]. A study by Charoenngam et al. [7] of variation orders on construction projects found that the average cost escalation was 7% of the original project cost with an average time extension of 30% than the originally stipulated project duration. Studies have also revealed that the significant reduction in both cost increase and time delay can be achieved through a complete design before commencement of work on site resulting in the prevention of variation orders [10].

Variation orders not only affect project performance in terms of time and cost but also adversely affect the quality, health, safety, and professional relations [5]. However, factors influencing the occurrence of variation orders and their adverse impact on project performance vary from one project to another. Factors include the nature of work, the complexity of the project, and the procurement method. Though it is likely that variation orders cannot be avoided completely, they can be minimized or prevented if their origin and causes were clearly known [6]. The successful execution of construction projects and keeping them within the estimated cost and prescribed schedules depend on methodology that requires sound engineering judgment [11].

One of the most important problems in the road construction industry in Sri Lanka seems to be managing variation orders. There is less literature available on this area; however, Pathiranage and Halwatura [12] have found in their study that one of the main causes of project delay is variation orders. Moreover, Nowfal and Gunawardana [13] stated that the variation order is a key factor which is contributing to price overrun in this industry.

This present study attempted to reveal the possible causes of variation orders in road construction projects in Sri Lanka. As the data collection was done through case study analysis and questionnaire survey, the accuracy of the data in the Sri Lankan context depends on the quality of the responses/respondents.

The main objectives of the study were as follows:(1)to find the causes of variation orders in the road construction industry in the international context through a literature review,(2)to find the causes of variation orders in road projects in Sri Lanka using a case study analysis and rank them based on the number of occurrences and percentage variation for each cause with respect to the initial value without contingencies,(3)to validate the findings using a questionnaire survey administered to industry experts and then rank them using the relative importance index (RII).

# CONCEPTUAL FRAMEWORK AND METHODOLOGY

The preliminary data for this research was collected through a literature review and a case study analysis. Further, a questionnaire was administered to professionals in the road construction industry in Sri Lanka. The literature review was done through books, conference proceedings, the Internet, and construction management and engineering journals.

The case study analysis was carried out for eleven project documents from recently completed road projects in Sri Lanka. An unbiased random sample of causes for variation orders was studied in order to predict the nature of the population (general circumstances) using statistical inference. RII was compiled based on the results of the questionnaire survey.

# RESULTS AND DISCUSSION

An effective analysis of variations and variation orders requires a comprehensive understanding of the root causes of variations [14]. Causes of variation orders were identified by many researchers [2, 15–20]. Based on their findings, 55 causes of variation orders were identified. Then they were grouped under four categories: owner related variations, consultant related variations, contractor related variations, and other variations. These findings were used to develop the questionnaire.

In this study, 50 respondents participated representing distinct road construction projects. Their background was analyzed with respect to their education, the level of occupation, and the number of years of working experience. While Figure 1 shows the respondents' educational background, Figures 2 and 3highlight the occupation level and the experience, respectively.

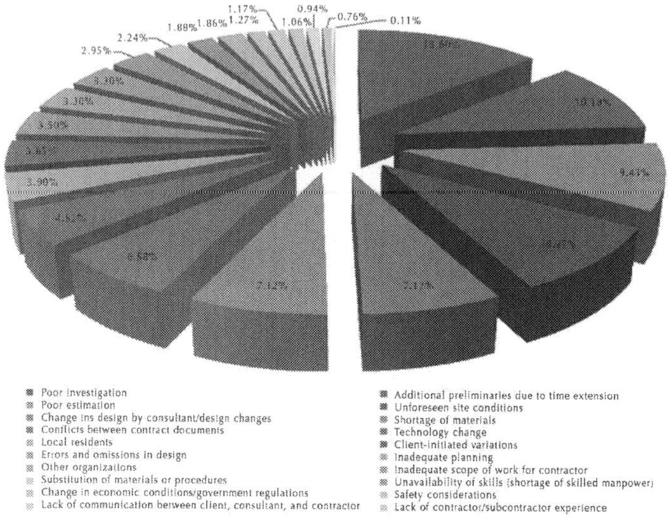


**Figure 6:** Contribution of each factor to variation.

These three separate findings are compiled into Table 1 together with the ranking found in the literature review.

**Table 1:** Comparison of ranks of each analysis

| Variation order cause | Number of occurrence | Percentage variation | Quest-ionnaire | Literature review |
|---|---|---|---|---|
| Poor estimation | 1 | 3 | 1 | — |
| Poor investigation | 2 | 1 | 4 | — |
| Unforeseen site conditions | 3 | 4 | 2 | 8 |
| Change in design by consultant/design changes | 4 | 5 | 11 | 5 |
| Additional preliminaries due to time extension | 5 | 2 | 24 | — |
| Client-initiated variations | 6 | 10 | 5 | 1 |
| Other organizations | 7 | 13 | 20 | — |
| Errors and omissions in design | 8 | 11 | 12 | 3 |
| Inadequate scope of work for contractor | 9 | 14 | 10 | — |
| Inadequate planning | 10 | 12 | 15 | 7 |

| | | | | |
|---|---|---|---|---|
| Local residents | 11 | 9 | 17 | — |
| Substitution of materials or procedures | 12 | 15 | 28 | 2 |
| Shortage of materials | 13 | 6 | 19 | — |
| Conflicts between contract documents | 14 | 7 | 30 | 6 |
| Technology change | 15 | 8 | 33 | 13 |
| Safety considerations | 16 | 18 | 32 | 12 |
| Unavailability of skills (shortage of skilled manpower) | 17 | 16 | 29 | 16 |
| Defective workmanship | 18 | 21 | 26 | 18 |
| Unrealistic contract durations imposed by client | 19 | 22 | 8 | — |
| Lack of contractor/ subcontractor experience | 20 | 20 | 14 | — |
| Lack of communication between client, consultant, and contractor | 21 | 19 | 25 | 7 |
| Change in economic conditions/government regulations | 22 | 17 | 9 | 19 |
| Political pressure | 23 | 23 | 3 | — |
| Owner's financial problems | — | — | 5 | 4 |
| The objective of the project is not well defined | — | — | 10 | 14 |
| The scope of work for the contractor is not well defined | — | — | 7 | 9 |
| Value engineering | — | — | 31 | 12 |
| Contractor's desire to improve his financial situation | — | — | 18 | 10 |
| The contractor's financial difficulties | — | — | 23 | 11 |
| Weather conditions | — | — | 13 | 17 |
| Natural disasters | — | — | 6 | — |
| Delay in approval | — | — | 16 | — |
| Consultant's lack of judgment and experience | — | — | 21 | — |
| Workmanship or material not meeting requirement of specifications | — | — | 22 | 15 |
| The required tools and equipment are not available | — | — | 27 | 18 |

According to the questionnaire survey, poor estimation was the most significant cause of variation orders. Unforeseen site condition was the second cause. Further, political pressure during construction stage, poor investigation, and client-initiated variations are at the 3rd, 4th, and 5th places in the ranking, respectively. According to the number of occurrence of the case study survey, poor estimation is the most significant factor. Poor investigation follows it. The rank of unforeseen site conditions is at the third place. Further, change in design by consultant/design changes and additional preliminaries due to time extension are at the 4th and 5th places in the ranking.

Based on the monitory value of the claims, poor investigation, additional preliminaries due to extension of time, poor estimation, unforeseen site conditions, and change in design by consultant/design changes are the first 5 causes, respectively.

The literature review suggests a slightly different scenario. According to that, client-initiated variations are the most significant factor. Further, change of plans by owner, substitution of material and procedure, errors and omission of design, owner's financial problems, and change in design by consultant are at 2–6, respectively, in the ranking.

It is apparent that the most significant factors that cause variation orders in the international context are not significant in the Sri Lankan context. Moreover, some causes which are not within the first twenty factors in the international context are more significant in the Sri Lankan context. For example, poor investigation and poor estimation are within the first four factors in the Sri Lankan context; however, internationally, they are not within the top 20 factors. According to the questionnaire survey results, political pressure experienced by professionals during the construction stage and owner financial problems are key factors for variation orders but they have not been documented.

Extension of time is considered a dominant factor for variation orders; however, professionals have not indicated it. Client seems to be a dominant factor for variation orders in both international and local contexts.

# CONCLUSIONS AND RECOMMENDATIONS

This study was focused on possible causes of variation orders in road construction projects in Sri Lanka. The data for the analysis were collected using three methods. The data related to causes of variation orders and their rankings were gathered based on a literature review. Then the findings were surveyed based on a questionnaire administered to the professionals in the road construction industry in Sri Lanka and several selected case studies were analyzed to understand the degree of occurrence and their impact.

This study identified that poor estimation is the main cause of variation orders in the road construction industry in Sri Lanka. For example, the consultants do not carry out adequate investigations at the initial investigation and design stage. Therefore, several site conditions rise in the construction stage. Further, clients do not have capable professional staff to carry out investigations and estimates, which are prepared by consultants and are not accurate in most cases. Therefore, several unnecessary variations occur during the construction stage. Apart from this main cause, most of the time, the fault of the client is not pointed out and not documented. Contractors and consultants are reluctant to point out client's faults, considering their survival in the construction industry. According to the questionnaire results, political pressure during construction stage is identified as a dominant factor. However, in case studies, they are not documented. Professionals do not seem to consider that the extension of allocated time is a critical factor for variations. However, according to the case studies, it is a critical factor.

Based on these findings it is possible to conclude that the Sri Lankan context should pay attention to several factors to minimize the occurrence of variation orders. The investigation must be carried out properly by qualified professional staff at the initial stage (in the pretender period) and adequate planning in advance is required by all involved parties before work starts at the site. The estimates have to be prepared properly by experienced professionals and clients should provide a clear brief of the scope of work. Further, consultants should ensure that the design/specifications fall within the approved budget and the budget team should be appointed early and they should participate in the designing process. Further, all parties should forecast unforeseen situations. Closer consultant coordination is required at the design stage, and utilization of an experienced consultant to produce a concluding design, working drawings, and contract drawings should be done at the tender stage. Further, maintaining proper communication channels and documentation is very important in terms of variation order management.

## REFERENCES


1. C. W. Ibbs, C. K. Wong, and Y. H. Kwak, "Project change management system," Journal of Management in Engineering, vol. 17, no. 3, pp. 159–165, 2001E. R. Fisk, Construction Project Administration, Prentice Hall, Upper Saddle River, NJ, USA, 5th edition, 1997.
2. J. J. O'Brien, Construction Change Orders, McGraw-Hill, New York, NY, USA, 1998.
3. FIDIC, "MDB Hardmonised Conditions of Contract for Construction—General Conditions," 2005.
4. F. M. Arain and L. S. Pheng, "How design consultants perceive causes of variation orders for institutional buildings in Singapore," Architectural Engineering and Design Management, vol. 1, no. 3, pp. 181–196, 2005.
5. A. A. Mohamed, Analysis and management of change orders for combined sewer over flow construction projects [Ph.D. Dissertation], Wayne State University, 2001.
6. C. Charoenngam, S. T. Coquinco, and B. H. W. Hadikusumo, "Web-based application for managing change orders in construction projects," Construction Innovation, vol. 3, pp. 197–215, 2003.
7. A. P. C. Chan and C. M. Yeong, "A comparison of strategies for reducing variations," Construction Management and Economics, vol. 13, no. 6, pp. 467–473, 1995.
8. F. M. Arain, "Strategic management of variation orders for institutional building: leveraging on information technology," in Proceedings of the PMI Global Congress Proceedings, pp. 1–17, Toronto, Canada, 2005.

9. P. A. Koushki, K. Al-Rashid, and N. Kartam, "Delays and cost increases in the construction of private residential projects in Kuwait," Construction Management and Economics, vol. 23, no. 3, pp. 285–294, 2005
10. A. H. Al-Momani, "Construction delay: a quantitative analysis," International Journal of Project Management, vol. 18, no. 1, pp. 51–59, 2000. View at Scopus
11. Y. L. Pathiranage and R. U. Halwatura, "Factors influencing the duration of road construction projects in Sri Lanka," ENGINEER Journal, vol. 43, no. 4, pp. 17–30, 2010.
12. M. A. M. Nowfal and N. D. Gunawardana, Study of contract Price Overunn with particular emphasis on claim situations in road and bridge contracts in the road development authority [M.S. thesis], University of Moratuwa, Katubedda, Sri Lanka, 1997.
13. W. Hester, J. A. Kuprenas, and T. C. Chang, "Construction changes and change orders: their magnitude and impact," CII Source Document 66, University of California-Berkeley, 1991.
14. CII, The Impact of Changes on Construction Cost and Schedule, Construction Industry Institute, University of Texas at Austin, Austin, Tex, USA, 1990.
15. H. R. Thomas and C. L. Napolitan, "Quantitative effects of construction changes on labor productivity," Journal of Construction Engineering and Management, vol. 121, no. 3, pp. 290–296, 1995
16. R. H. Clough and G. A. Sears, Construction Contracting, John Wiley & Sons, New York, NY, USA, 6th edition, 1994.
17. J. Wang, N. Fisher, C. Sun, and D. Wu, "An analysis of the distribution of time variance for building projects," The International Journal of Construction Management, vol. 3, no. 1, pp. 73–82, 2003.
18. Y. Wang, "Coordination issues in Chinese large building projects," Journal of Management in Engineering, vol. 16, no. 6, pp. 54–61, 2000
19. F. M. Arain and L. S. Pheng, "The potential effects of variation orders on institutional building projects," Facilities, vol. 23, no. 11-12, pp. 496–510, 2005

Chapter 2

# A Cost-Based Interior Design Decision Support System for Large-Scale Housing Projects

Hoon-ku Lee[1], Yoon-sun Lee[2] and Jae-jun Kim[3]

[1]LIG Engineering and Construction Co., Ltd.

[2]Department of Architectural Design, Hanyang University

[3]Department of Sustainable Architectural Engineering, Hanyang University Korea

## INTRODUCTION

In the early stages of a large-scale housing project, many interior design alternatives remain to be confirmed after a rough review of the costs. In general, interior designers consider the overall concept, colour and style according to floor plans, spaces and elements. However, because they generally do not consider the construction work required, the construction and design characteristics are not connected, and the cost and design properties are controlled separately. This is why the real-time management of cost change is not included in the decision-making process. It is therefore necessary to consider the cost when making decisions on interior design items for an apartment unit plan (Lee et al., 2007). Rapid advances in information technology are changing the nature of most human activities (Bennett, 2000) and are generating new requirements for clients such as the owner/developer in large-scale housing projects. The clients' requirements are stated in ongoing communications among project participants in the early phase of a project and become embodied in the design phase, during which design alternatives must be selected to meet the clients' requirements while satisfying them in a realistic way. A decision support system for selecting design alternatives is intended to represent design information, document design rationale and manage design changes (Geoffrion, 1987). Many studies on the design phase have focused on the cooperation between various participants such as the architects, engineers and contractors (A/E/C) (Demirkan, 2005; Kalay et al., 1998; Khedro et al., 1994; Lee et al., 2001; Mokhtar et al., 2000), but few studies have examined cooperative systems or decision support systems in which the end-user or client participates in the selection of interior design specifications. In addition, it is unusual to propose a process or database function that accepts and manages the extensive interior design information generated by too many alternatives. Integrated project systems would help streamline project activities by allowing downstream disciplines to access design information. With this, they could evaluate the design and assess the impact of design decisions on downstream project activities early in the design process (Halfawy & Froese, 2005). In this chapter, we aimed to devise a system that allows clients to make cost-based decisions suited to their own interior design specifications and that enables the builder to plan resource requirements and budget costs. We describe an information model that supports cost-based decision making in the interior design phase. To do this, we derive the space hierarchy for a large-scale housing project. We also propose a method for building a library of interior design information based on the space hierarchy and interior object information. The proposed model is validated using an example study analysis to show how it supports the decision-making process of various participants in the interior design phase by providing real-time cost information when the interior is initially planned or later changed.

# A CONCEPT OF COST-BASED INTERIOR DESIGN

## Current interior design procedure

The interior design progresses through the stages of conception, modelling, review, finalisation, detailing, drafting and costing as shown in Fig. 1. Interior designers working on projects come up with design concepts, perform space modelling, review results, fill in details of the finalised design and draft the results, all to calculate the costs for estimates.

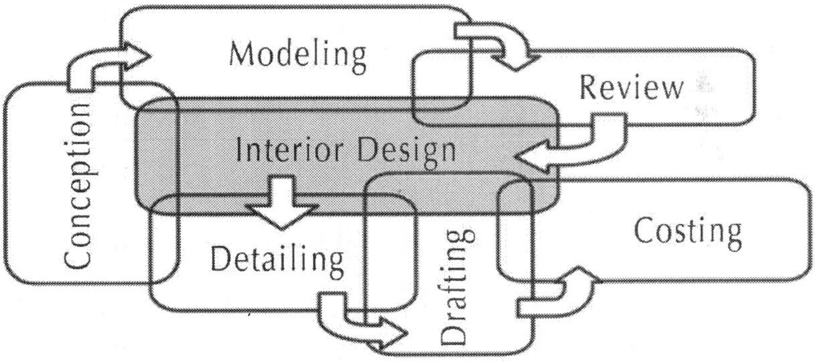


**Figure 1:** Interior design phases.

In the existing interior design procedure shown in Fig. 2, owners/developers first produce ideas or requests. Then interior designers propose designs and alternatives, and the results are initially reviewed without any consideration of cost. In the detailed design stage, the owner or developer reviews the interior design along with costs.

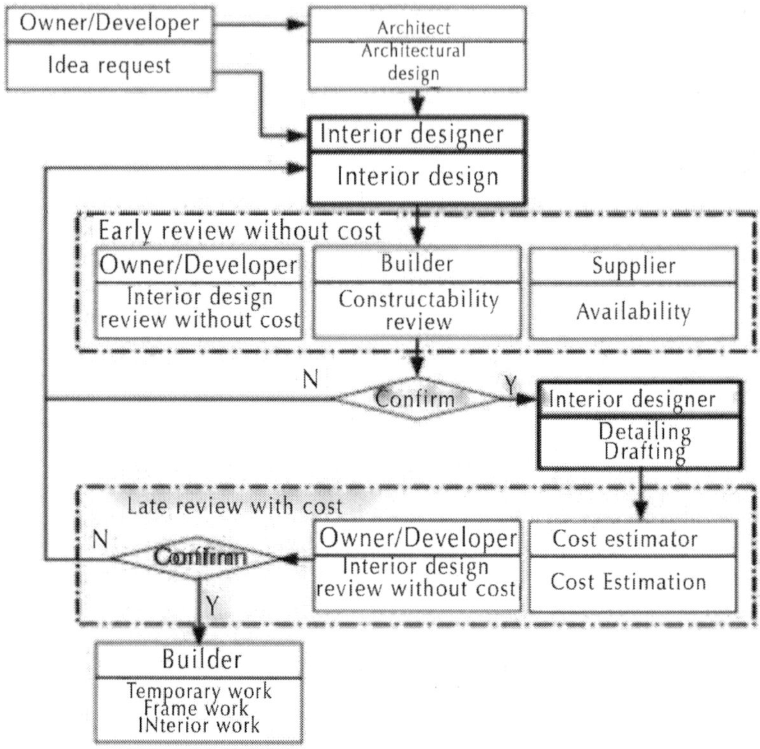


**Figure 2:** Current interior design procedure.

However, the procedure is a long one, and is subject to many problems in feedback concerning alternative design proposals. During the design stage, design alternatives for finishing materials are based on the planned space and rooms in a unit, according to the demands of the project manager; these alternatives are then incorporated into the unit plan. This process usually generates a good amount of cost information, although it usually fails to manage cost information, one of the fundamental criteria of project performance. Design characteristics and costs are generally not assessed simultaneously as finishing materials are selected because the process is managed on a basis of dualization. Therefore, no cost baseline is determined at each project stage. The interior designer usually does not consider costs generated by the selection of finishing materials, focusing instead on design characteristics such as the concept, color, and pattern for the space or unit plan.

## Proposed Cost-Based Interior Design Procedure

Recently, experts have started to consider a design-to-cost philosophy to be a necessary requirement for effective project cost management. The design-to-cost method can produce accurate estimates of the cost to produce products or services before the project begins, systematically constraining design goals based on available funds (Michaels & Wood, 1989). Thus, the design-to-cost management strategy and supporting methodologies can achieve an affordable product by treating target cost as an independent design parameter that needs to be achieved during project development. Achieving highly cost-effective results requires assessing costs related to various approaches and design solutions. Parametric cost estimation models have been developed (Kim et al., 2004). Regression, or multiple regression analysis as it is usually called, is a very powerful statistical tool that can be used as both an analytical and predictive technique for examining the contribution of potential new items to the overall cost estimate reliability (Hegazy et al., 2001). It is not appropriate, however, when describing nonlinear or multidimensional relationships with multiple inputs and outputs (Huyn et al., 1993). In addition, it is difficult to use parametric methods when the number of alternatives tends be infinite, based on the different items of the design.

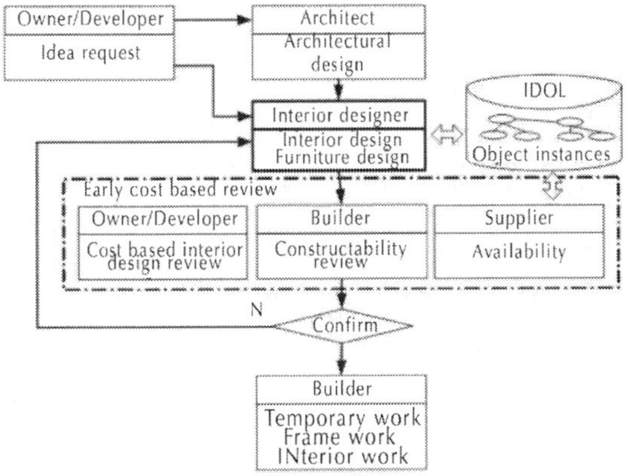


**Figure 3:** Proposed cost-based interior design procedure.

Therefore, we propose a procedure in which designers use an interior design object library (IDOL) to select an interior design item based on cost. At that point, the total cost for the interior can be reviewed. The builder's constructability review and the suppliers' availability review of the interior design are not included in this study, but the design information stored in the interior design information model can be used in the construction phase.

# A MODEL FOR REPRESENTING INTERIOR DESIGN INFORMATION

Various approaches have been proposed to provide structure to product models. Early efforts included the A/E/C building systems model (Turner, 1990) and the general A/E/C reference model (GARM) (Gielingh, 1998). The major standardisation effort in product modelling today is ISO-STEP from the International Standards Organization (Hegazy et al., 2001), and in recent years, researchers in the A/E/C industry have devoted considerable attention to the representation of design information and the management of design changes. Researchers and practitioners have been investigating improved integration, that is, the continuous and interdisciplinary sharing of data, knowledge and goals among all project participants (Hegazy et al., 2001; Luiten & Tolman, 1997). The architectural information model proposed in these studies manages the creation, modification and exchange of the spatial design information created for all participants in the process for A/E/C purposes (Hegazy et al., 2001). Other studies have proposed building information models from various different perspectives. Anwar (2005) studied methods of structural analysis, modelling and design with structural mechanisms for major members (e.g., foundation, column, beam, wall, slab) in building structures using a structural information model. Choi et al. (2007) provided a building data model including building components such as building, plan, space, ring, wall skeleton, surface and column, for structured floor plans. Figure 4 shows the building project hierarchy (BPH) of a building information model.

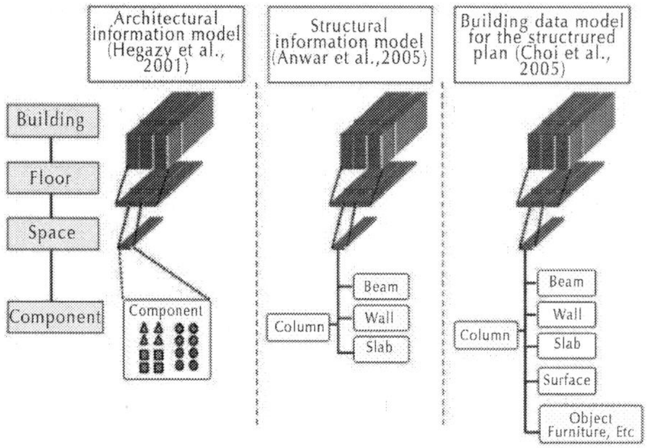


**Figure 4:** Building project hierarchy according to the building information model.

In this study, the structured floor plan (Choi et al., 2007) was used as the basis of the interior design information for housing projects. The structured floor plan is roughly divided into private space and public space. Its major components are surfaces such as floors, walls and ceilings, and non-surfaces such as furniture, windows, doors and lighting fixtures. These are the major components of private space for an apartment unit plan and serve as the primary focus of this study.

# A COST-BASED INTERIOR DESIGN DECISION SUPPORT SYSTEM

## Roles of an Interior Design Decision Support System

### *Communicating With Clients in Deciding Interior Design Specifications*

The evaluation of design alternatives is an important ongoing phase in the design decision stage and when producing new design concepts.

Most design decisions are made intuitively without predicting the actual performance with respect to a variety of parameters such as lighting, energy and comfort (Reichard & Papamichael, 2005). In addition, design decisions are made by expert project participants without regard to the requirements of the occupants in terms of space, element colour, pattern, texture and materials. Consequently, the clients frequently change the interior design when they review the feasibility study and plan the target cost. The added cost, time delay and construction waste generated by such changes lead to claims and waste resources. Therefore, a decision support system must allow client input when the interior design specifications are selected. The owner/developer's requirements lead to design alternatives for each room, and the alternatives are incorporated in space (i.e., the unit plan) in the interior design stage. Also present at this stage is a deluge of cost information, which makes it difficult to generate and manage the costs that predominate in project. Such a problem is also linked to the communication difficulties among the project participants such as the owner/developer, interior designer and builder.

## *Providing a Cost Baseline when Selecting Interior Design Specifications*

Project cost management deals with the procedures to ensure that the project is completed within the approved budget. The Project Management Body of Knowledge explains project cost management as a four-phase process consisting of resource planning, cost estimation, budget establishment and cost control (PMI, 2000). By linking this with the project work breakdown structure (WBS), we propose a cost baseline to control costs during the design phase, which enables the builder to examine resource and cost planning during the interior design phase of a project. Our system reports the changes that a client makes to the interior design, which affect resource planning in the WBS and alter the costs.

## *Managing an IDOL for Integrating Design and Cost Information*

To control costs during the interior design phase, interior design information, process characteristics, and cost information must be interrelated. To control the information on the finishes work item generated through the interrelation of the construction and design characteristics, we developed an IDOL. We used a relationship analysis to examine several completed apartment projects to identify interior design objects. The relationships involve spaces, rooms, components (surfaces and non-surfaces), work items, and design ï cost ï work information. A work item is defined as an interior design object that is integrated with the proposed BPH. For work items, the IDOL includes the surfaces and Non-surfaces design information (colour, material, pattern, texture, image), cost information (quantity, unit, unit price, cost), and work information (specification, size, cad file, work breakdown structure), as shown in Fig. 5. Surfaces include floors, walls and ceilings, while non-surfaces include furniture, windows, doors and lighting fixtures. These are the major components of private space for an apartment unit plan and serve as the primary focus of this study.

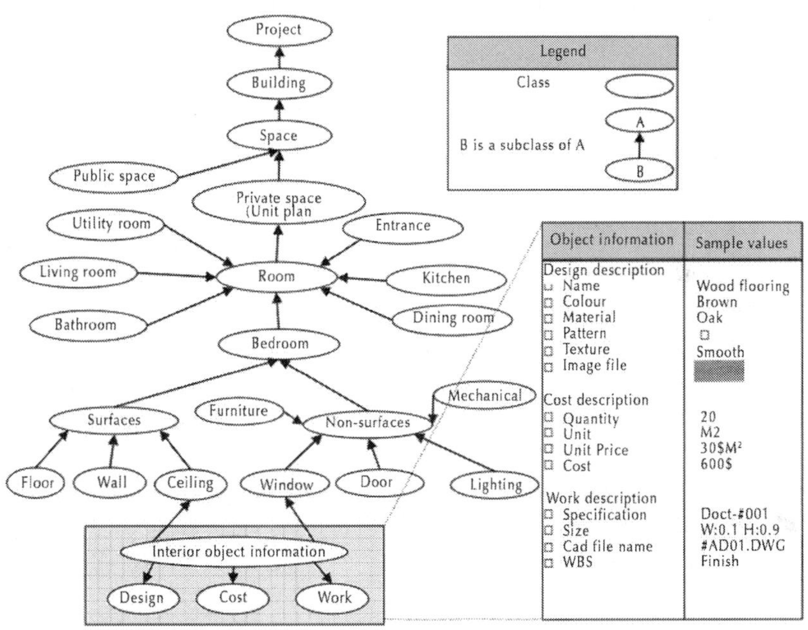


**Figure 5:** Hierarchy schematic for interior object information.

Once the information is integrated, it can be used by each project participant for his or her particular responsibilities as Fig. 6 suggests.

For example, the interior designer reviews the costs and plans several different design alternatives for each room based on the unit plan. The owner/developer can review the project costs for each interior design alternative, and the builder will be able to prepare resource and budget plans for each WBS.

## Schematic of the Interior Design Decision Support System

The Interior Design Decision Support System suggested here is capable of controlling interior design costs and supporting client decisions. This is achieved by referring to resource consumption plans from the early phases of a project, and by implementing and using the IDOL database to link the construction and design characteristics for interior design. The procedure involved is shown in Fig. 7, and the components of the system are described in the following section.

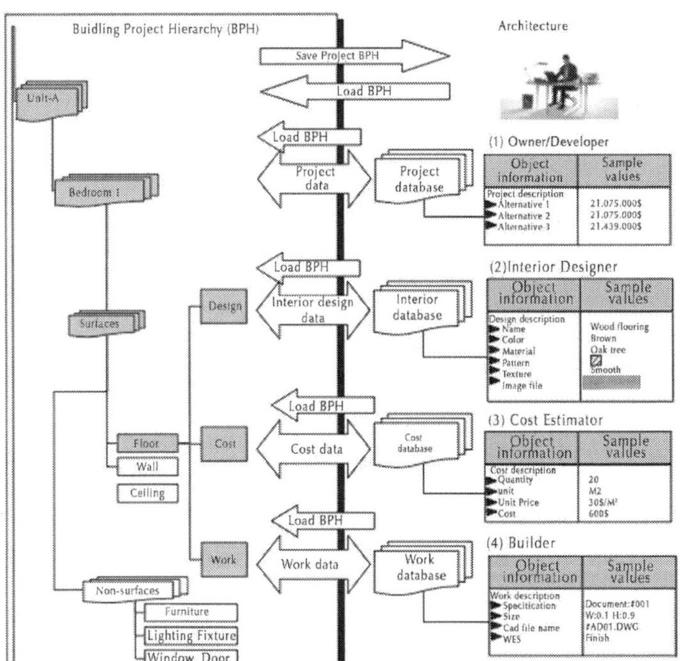


**Figure 6:** Multidisciplinary use of the BPH and the IDOL.

# COMPONENTS OF THE INTERIOR DESIGN DECISION SUPPORT SYSTEM

## Users

The main users of the Interior Design Decision Support System are the client, the interior designer and the builder, who participate in the decision-making process. The interior designer has the responsibility for the interior design process. Once the interior designer selects the default values for the IDOL, clients can change the colour, pattern, texture and material by viewing images of the interior design objects. They can also examine costs, one of the unit measures, on a real-time basis. As shown in Fig. 7, after receiving a BPH from an expert group involving A/E/C, the interior designers select the default values for the IDOL. Then clients are given the task of making selections from the IDOL. With help from a cost estimator, construction costs are estimated for the selected interior design, and in addition, the particulars of the changes in the IDOL are updated and controlled.

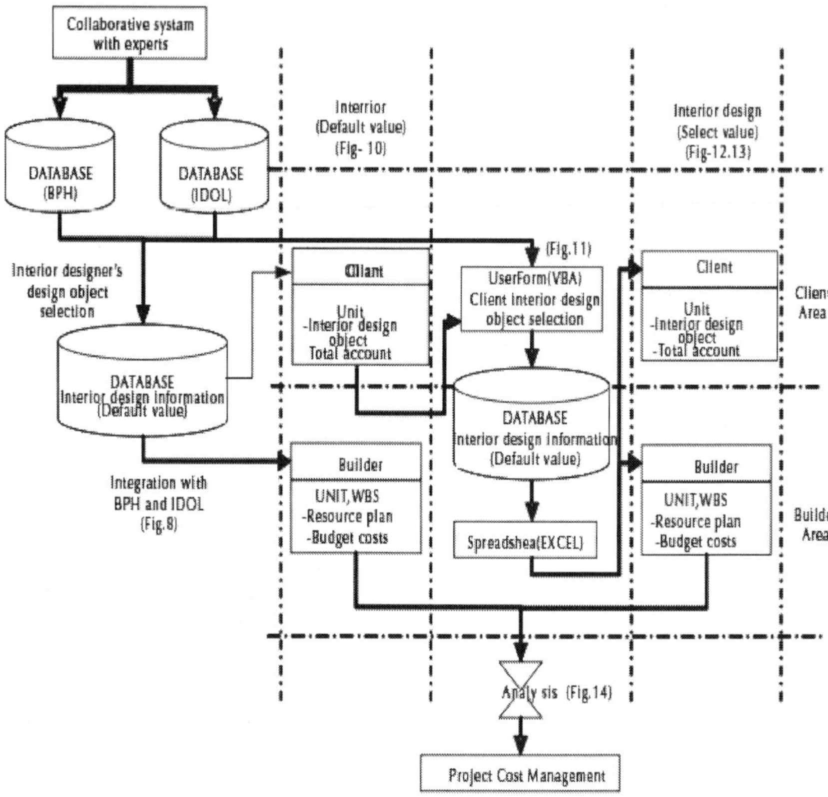


**Figure 7:** Schematic of the proposed Interior Design Decision Support System.

## Decision-making Procedure

After the A/E/C group produces the BPH during the detailed design phase, interior designers select the design objects from the IDOL and set the default values for the interior design. Targeting these default values, the clients use the decision support system and make decisions by selecting interior design objects to their requirement from the IDOL. Clients can make decisions either under the constraint of fixed costs (e.g., simply changing the colour of wallpaper), or with the option of variable costs to change the design items.

## ***Information Model***

The underlying data of the Interior Design Decision Support System are related to the work involved and the costs of the interior specifications. As stated previously, the information on interior design decisions comes from the creation of the IDOL and its integration with the project BPH. As shown in Fig. 8, the proposed BPH and the IDOL produce interior design information that is integrated after being selected by the client. Here, the IDOL holds the elements such as work items, specification, image, colour, material, unit, unit price and work breakdown structure. The interior design information (the selected value) generated here is displayed and used to meet the requirements of the client and builder.

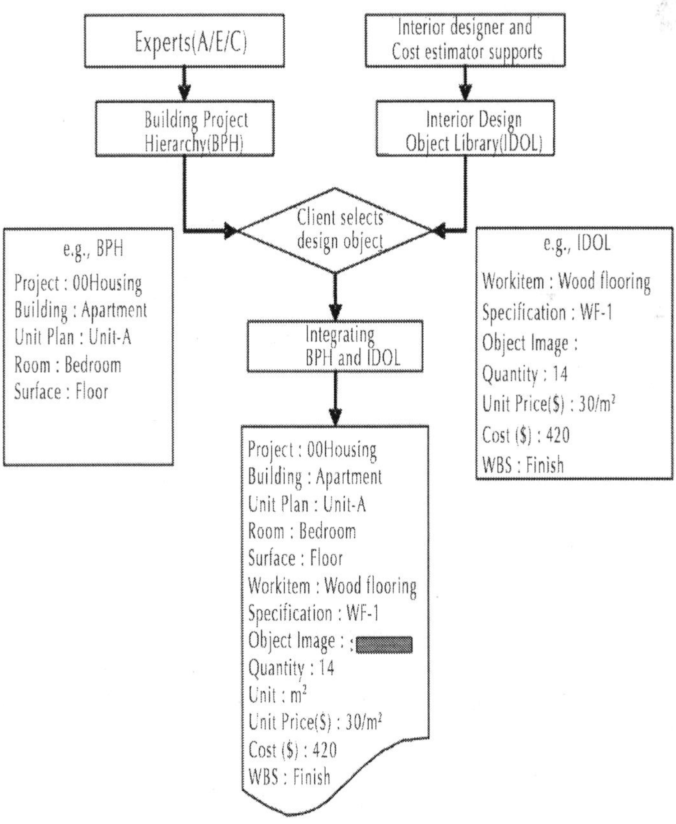


**Figure8:** Integrating the BPH and the IDOL

As seen in Fig. 9, an information model was suggested to integrate the BPH and IDOL that supports cost-based interior design decisions.

# EXAMPLE STUDY AND APPLICATION

## Establishing the Interior Design Information (Default Values)

To set the values for the interior design objects and costs in the early phase, the BPH of the target project is configured in co-operation with the A/E/C experts, and the default values of interior design objects are created with the help of the designer and cost estimator. The cost of each unit is established when the interior designer inputs the interior design for the units using design objects from the BPH, and the cost estimator inputs the cost estimates for the selected interior design. Figure 10 shows an example of the default values set by the interior designer for each element.

## The Client Selects the Interior Design Objects

The client makes decisions with the support of the IDOL database, which is linked to the default values. As design objects are selected, the cost is changed accordingly. Using the basic default values, the client selects the interior specifications that their requirements. Two scenarios for changes can be simulated involving the fixed and variable costs.

### *Fixed Cost Example*

When the client selects the fixed cost condition, he or she can decision making the alternatives provided in the IDOL database, as presented in Fig. 11. The results of these decisions are displayed on a spreadsheet, as shown in Fig. 12. For example, when the client changes the wallpaper pattern, the interior design objects change, although no change in the cost is indicated. Nevertheless, the client can still choose from a variety of alternatives. The information generated in this example is then used in resource planning and project cost management.

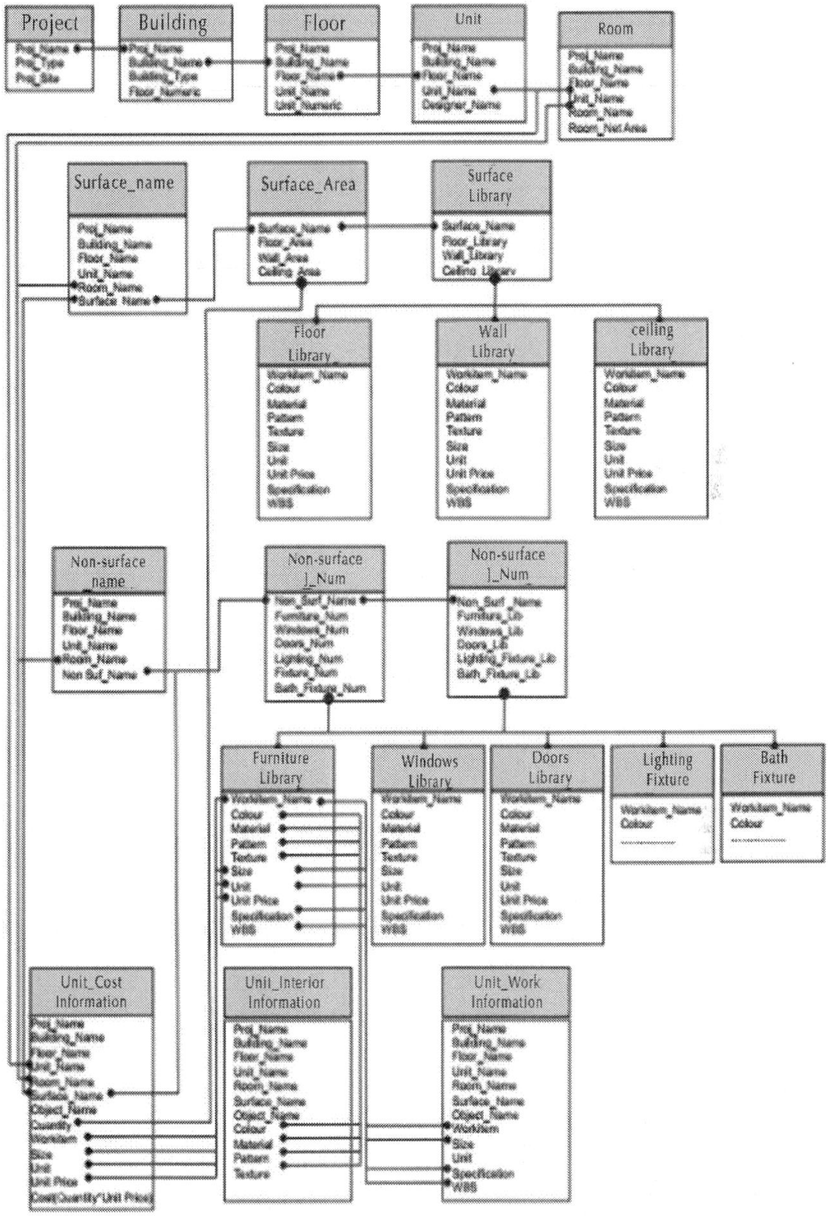


**Figure 9:** Interior design information model.

| Default (1) | Information | Floor | Wall | Ceiling | Decoration | Doors | Fixture | Total |
|---|---|---|---|---|---|---|---|---|
| 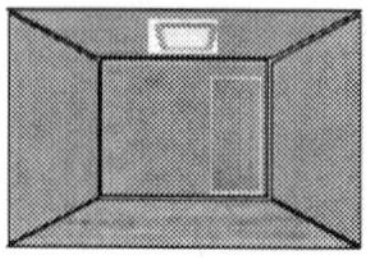 | Interior work item | Wood flooring | Wallpaper | Wallpaper | Moulding | Wood door | Lighting fixture | |
| | Specification | WF-1 | WP-1 | WP-1 | MD-1 | 0.9*2.0 | LF-01 | |
| | Object Image | | | | | | | |
| | Quantity | 14 | 28 | 14 | 15 | 1 | 1 | |
| | Unit | $M^2$ | $M^2$ | $M^2$ | M | EA | EA | |
| | Unit Price ($) | 30 | 5 | 5 | 3 | 250 | 150 | |
| | Cost ($) | 420 | 140 | 70 | 45 | 250 | 150 | 1,075 |
| | WBS | Finish | Finish | Finish | Decoration | Doors | Electnic | |

**Figure 10:** Interior designer-selected values (default values)

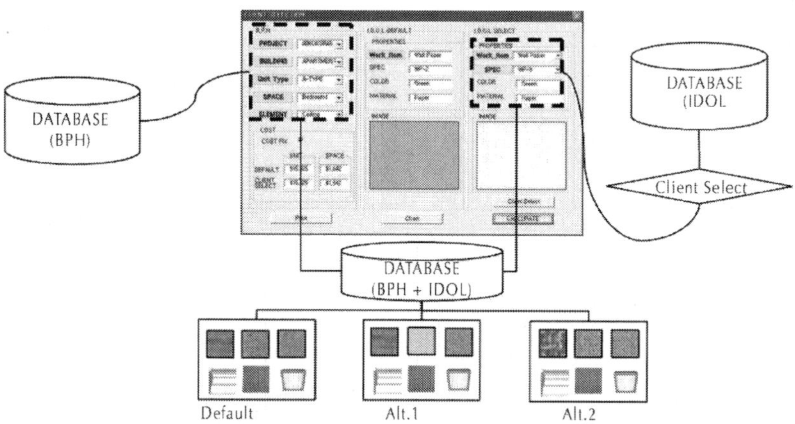


**Figure 11:** Decision making the alternatives in IDOL database.

| Alternative 1 | Components | Floor | wall | ceining | Decoration | Doors | Fixture | Total |
|---|---|---|---|---|---|---|---|---|
| 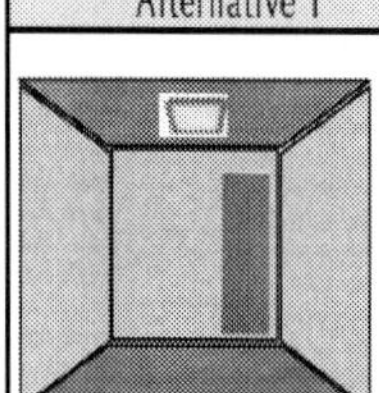 | Interior work item | Wood flooring | Wallpaper | Wallpaper | Moulding | Wood door | Lighting fixture | |
| | Specification | WF-1 | WP-2 | WP-1 | MD-1 | 0.9*2.0 | LF-01 | |
| | Object Image | | | | | | | |
| | Quantity | 14 | 28 | 14 | 14 | 1 | 1 | |
| | Unit | $M^2$ | $M^2$ | $M^2$ | M | EA | EA | |
| | Unit Price($) | 30 | 5 | 5 | 3 | 250 | 150 | |
| | Cost ($) | 420 | 140 | 70 | 45 | 250 | 150 | 1,075 |
| | WBS | Finish | Finish | Finish | Decoration | Doors | Electric | |

**Figure 12:** Options for the client-selected values under the fixed cost scenario.

### *Variable Cost Example*

Alternatively, the client can select finishes and confirm the final costs on a real-time basis. These decisions are made by selecting from alternatives proposed in the IDOL database when selecting the variable cost option. As Figure 13 shows, when the client changes work items for the floor, wall and ceiling, the cost changes from $1,075 to $1,439. A system example study was run using the default values proposed by the interior designer using the fixed and variable cost options. The change in cost is the difference between the cost of the interior design objects proposed by the interior designer, and the cost of those selected by the client. Under the fixed cost scenario, the client changed the design of wall objects without changing the costs relative to the default values set by the interior designer. With the variable cost option, changing the wood flooring to tile carpet and the wallpaper material from paper to fabric increased the cost to $1,439 from the default value of $1,075.

| Alternative 1 | Components | Floor | wall | ceining | Decoration | Doors | Fixture | Total |
|---|---|---|---|---|---|---|---|---|
| 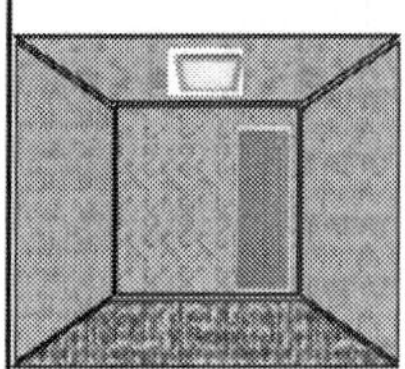 | Interior work item | Carfet tile | Wallpaper | Wallpaper | Moulding | Wood door | Lighting fixture | |
| | Specification | CT-2 | WP-3 | WP-3 | MD-1 | 0.9*2.0 | LF-01 | |
| | Object Image | | | | | | | |
| | Quantity | 14 | 28 | 14 | 15 | 1 | 1 | |
| | Unit | $M^2$ | $M^2$ | $M^2$ | M | EA | EA | |
| | Unit Price ($) | 50 | 7 | 7 | 3 | 250 | 150 | |
| | Cost ($) | 700 | 196 | 98 | 45 | 250 | 150 | 1,439 |
| | WBS | Finish | Finish | Finish | Decoration | Doors | Electric | |

**Figure 13:** Options for the client-selected values under the variable cost scenario.

## Application for the Client and Builder

Considering the needs of the client/builder described in Section 4.1, the interior design objects selected for the target project are integrated for each space and WBS, and the results are displayed for the client and builder. When examining the costs, the client can also check the interior design total project cost by decisions made concerning the unit interior design. The builder can make a project interior construction cost plan and resource consumption plan according to the WBS. The following detailed summaries are produced for the client and builder.

## Application for the Client

As shown in Fig. 14, the developer or owner can establish a baseline cost of \$21,075,000 against which to compare the alternative proposals. Alternative 1 has no cost impact compared to the baseline, while Alternative 2 would increase the project cost by \$364,000 due to the interior changes.

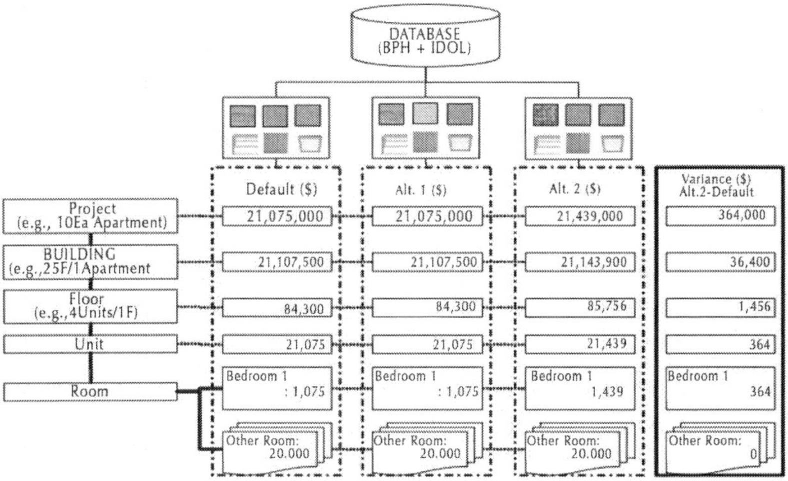


**Figure14:** The total costs according to the changes

## Application for the Builder

Based on the client's decisions, the builder is provided with cost and resource data according to the WBS (e.g., finishes, doors, windows and furniture), displayed as the amount per construction type as shown in Fig. 15(a). The builder or other agents can also use work items from the spaces as shown in Fig. 15(b).

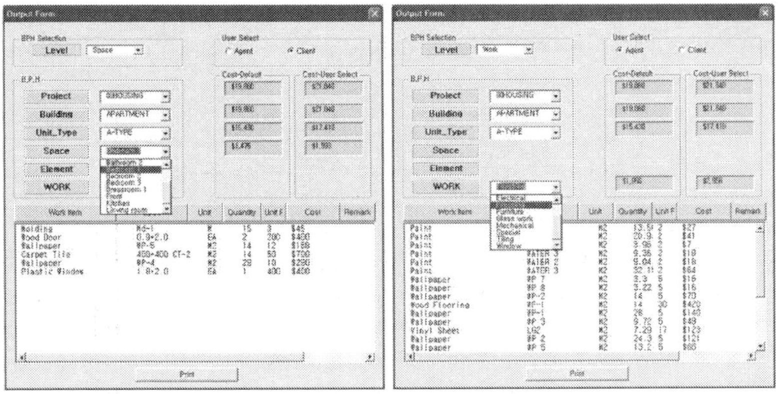


(a) According to WBS (b) According to space resources

**Figure 15:** The user interface of default interior design and client-selected items.

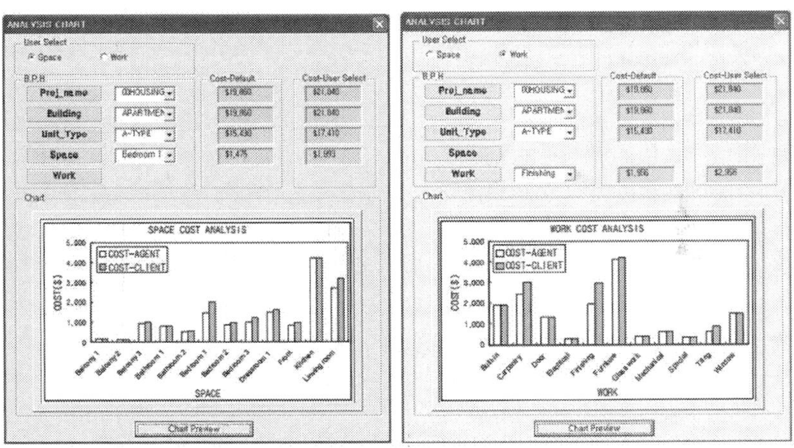


(a) According to space (b) According to works

**Figure 16:** Cost analysis charts

The design objects and costs generated here are used as data for managing project costs in the WBS as shown in Fig. 16. By analysing the costs according to the different spaces (Fig. 16(a)) and work (Fig. 16(b)), the builder can make a construction cost plan and resource consumption plan in accordance with the WBS. From the perspective of project cost management, the default cost proposed by the interior designer and the costs determined by client selection can increase the estimate accuracy of the project costs in the early phases of a project.

# CONCLUSION

We proposed an interior design decision support system that enables clients to participate in the early phases of the interior design for a housing project. This allows them to select objects (e.g., colours, patterns, material) for the interior design based on the unit measure of costs with input from various A/E/C experts. This also allows the interior designer and cost estimator to check the costs accordingly. The proposed system was implemented using the BPH and IDOL. In an example study, interior design objects and default costs were proposed while examining the fixed and variable cost options. This study examined how client decisions made during the early design phases and based on restrictions in the interior design can affect a project. An analysis of the results confirmed that our interior design decision support system leads to client satisfaction with the interior design, while enabling clients to manage project costs by providing a cost baseline. Overall, this increases the accuracy of early estimates made during the project concept phase. Cost and time are both very key indicators for assessing project performance. In the earlier phases of large-scale housing projects, the review of construction costs is important (Kim et al., 2004). It is not appropriate to state that high-cost interiors designs are good and low-cost ones are not. However, making interior designs or design changes without any consideration of the cost is a source of serious problems for the interior designer and other project participants, especially the owner/developer. Moreover, despite considerable information on costs for alternative proposals selected by interior designers, existing procedures have many problems as seen in a review and feedback of these alternatives. To address these problems, we investigated various alternative proposals for surfaces (floor, wall and ceiling) and non-surfaces (furniture, windows and doors) on a unit plan basis for large-scale housing development projects, and provided an interior information model. The model was validated through an example study to show how it could be used in the decision-making process by various participants in a construction project. The proposed model is useful in providing a total interior cost review and cost baseline for the developer/owner, a means for cost-based decision making by interior designers, and the interior material information required by builders. Future studies will focus on the automation of quantity surveying and graphics to reinforce the use of Internet-based decision support systems. In future studies, further subdivision of the design objects characteristics will be necessary, as well as the development of a system that enables clients to make interior design decisions by changing layers, a subject outside the scope of the present study. Additional work is also required to enable interior designers to use the information provided by vendors and suppliers directly, and to provide the information for the procurement phase without any additional processing.

# REFERENCES


1. Anwar, N. (2005). Component-based, information oriented structural engineering applications.Journal of Computing in Civil Engineering, Vol. 29 (1), pp 45–57.
2. Bennett, J. (2000). Construction the third way: managing cooperation and competition in
3. construction, Butterworth–Heinemann, Boston, MA.
4. Björk, B.C. (1989). Basic structure of a proposed building product model. Computer Aided
5. Design, Vol. 21 (2) pp 71–78.
6. Choi, J.W., Kwon, D.Y., Hwang, J.E., & Lertlakkhanakul, J. (2007). Real-time management of spatial information of design. Automation in Construction, Vol. 16 (4), pp 449–459.
7. Dawood, N., Sriprasert, E., Mallasi, Z., & Hobbs, B. (2003). Development of an integrated
8. information resource base for 4D/VR construction processes simulation. Automation in
9. Construction, Vol. 12 (2), pp 123–131.
10. Demirkan, H. (2005). Generating design activities through sketches in multi-agent system.
11. Automation in Construction, Vol. 14 (6), pp 699–706.

12. Geoffrion, A.M. (1987). An introduction to structured modelling. Management Science, Vol. 33,pp 547–588.
13. Gielingh, W. (1988). General AEC Reference Model. ISO TC184/ SC4/WG1 doc. 3.2.2.1, ISO,Delft, The Netherlands.
14. Halfawy, M. & Froese, T. (2005). Building integrated architecture/ engineering/construction systems using smart objects. Journal of Computing in Civil Engineering, Vol. 19 (2), pp 172–181.
15. Hegazy, T., Zaneldin, E., & Grierson, D. (2001). Improving design coordination for building projects. Journal of Construction Engineering and Management, Vol. 127 (4), pp
16. 322–329.
17. Huyn, P.N., Geneserth, M.R., & Letsinger, R. (1993). Automated concurrent engineering in design. World Computing, Vol. 26 (1), pp 74–76.
18. ISO (1994). ISO 10303-1 Part 1: Overview and fundamental principles, International Organization for Standardization, Geneva, Switzerland.
19. Kalay, Y.E., Khemluni, L., & Choi, J.W. (1998). An integrated model to support distributed collaborative design of buildings. Automation in Construction, Vol. 7 (2–3), pp 177–188.
20. Khedro, T., Teicholz, P., & Geneserth, M.R. (1994). A framework for collaborative distributed facility engineering. Proceedings of the 1st Congress of Computing in Civil Engineering, ASCE, New York, NY, pp 1489–1496.
21. Kim, G.H., An, S.H., & Kang, K.I. (2004). Comparison of construction cost estimating models based on regression analysis, neural networks, and case-based reasoning. Building and Environment, Vol. 39 (10), pp 1235–1242.
22. Korean National Statistical Office, <http://www.nso.go.kr/>
23. Lee, E.J., Woo, S.G., & Sasada, T. (2001). The evaluation system for design alternatives in collaborative design. Automation in Construction, Vol. 10 (3), pp 295–301.
24. Lee, H.K., Lee, Y.S., Kim, K.H., & Kim, J.J. (2007). A Cost-based Information Model for an Interior Design in a Large-scale Housing Project, ICCIT 07, 2007 International Conference on Convergence Information Technology.

25. Luiten, G.T.B. & Tolman, F.P. (1997). Automating communication, in civil engineering. Journal of Construction Engineering and Management, Vol. 123 (2), pp 113–120.
26. Mokhtar, A., Beard, C., & Fazio, P. (2000). Collaborative planning and scheduling of interrelated design changes. Journal of Architectural Engineering, Vol. 6 (2), pp 66–75.
27. Project Management Institute (2000). A Guide to the Project Management Body of Knowledge (PMBOK® Guide). Project Management Institute, Singapore.
28. Reichard, G. & Papamichael, K. (2005). Decision-making through performance simulation and code compliance from the early schematic phase of building design. Automation in
29. Construction, Vol. 14 (2), pp 173–180.
30. Turner, J.A. (1990). AEC Building Systems Model. ISO TC184/SC4/WG1 doc. 3.2.2.4, ISO, Delft, The Netherlands.

Chapter 3

# Benefits of Building Information Modelling in the Project Lifecycle: Construction Projects in Asia Regular Paper

Jian Li[1], Ying Wang[2], Xiangyu Wang[2,3], Hanbin Luo[4], Shih-Chung Kang[5], Jun Wang[2], Jun Guo[6] and Yi Jiao[7]

[1]Department of Civil Engineering, East China Jiaotong University, China

[2]Australasian Joint Research Centre for BIM, School of Built Environment, Curtin University, Australia

[3]Department of Housing and Interior Design, Kyung Hee University, South Korea

[4] School of Civil Engineering & Mechanics, Huazhong University of Science and Technology, China

[5]Department of Civil Engineering, National Taiwan University, Taiwan

[6]CCDI Group, China

[7]School of Computer Science, Fudan University, China

# ABSTRACT

Building Information Modelling (BIM) is a process involving the creation and management of objective data with property, unique identity and relationship. In the Architecture, Engineering and Construction (AEC) industry, BIM is adopted a lot in the lifecycle of buildings because of the high integration of information that it enables. Four-dimensional (4D) computer-aided design (CAD) has been adopted for many years to improve the construction planning process. BIM is adopted throughout buildings' lifecycles, in design, construction and operation. This paper presents five large-scale public and financial projects that adopt BIM in the design, construction and operational phases. Different uses of BIM are compared and contrasted in the context of the separate backgrounds. It is concluded that productivity is improved where BIM is used to enable easy sharing and integration of information and convenient collaboration. Keywords Building Information Modelling, 4D, Construction, Building Lifecycle

# INTRODUCTION

Building Information Modelling (BIM), as a powerful set of tools, has been widely applied in the Architecture, Engineering, and Construction (AEC) industry. The inherent nature of AEC makes it a suitable application domain for virtual reality, particularly for design and planning [1]. Schlueter and Thesseling defined three major types of information involved in BIM – geometric, semantic and topological [2]. Geometric information means the 3D modelling of a building, and semantic information includes the properties of components. Topological information describes the dependency relationship between properties and components. A BIM model could comprise individual 3D models of each building component, with all associated properties such as weight, material, length, height, etc. [3]. Beyond the inherent information, BIM also includes external associations between building components. BIM has been applied in the areas of structure, energy, disaster prevention, construction planning and scheduling, project control, construction safety, and maintenance [4]. BIM has also been implemented with Augmented Reality (AR) to improve the productiveness of on-site work [5]. BIM is a revolutionary technology that is transforming the way buildings are designed, planned, constructed and managed [6]. This paper will present and analyse five BIM projects in terms of building lifecycles. The adoption of BIM is at different stages within different backgrounds; the methodology of analysing, comparing and contrasting the cases is therefore presented in the following section.

# METHODOLOGY

This paper will present and study five construction projects that have adopted BIM in different phases of the building lifecycle in various circumstances, and ask how BIM is adopted in the design, construction and operational phases. Since these projects have different backgrounds, a case study structure is developed to allow each case to be analysed and compared in different aspects (Figure 1).

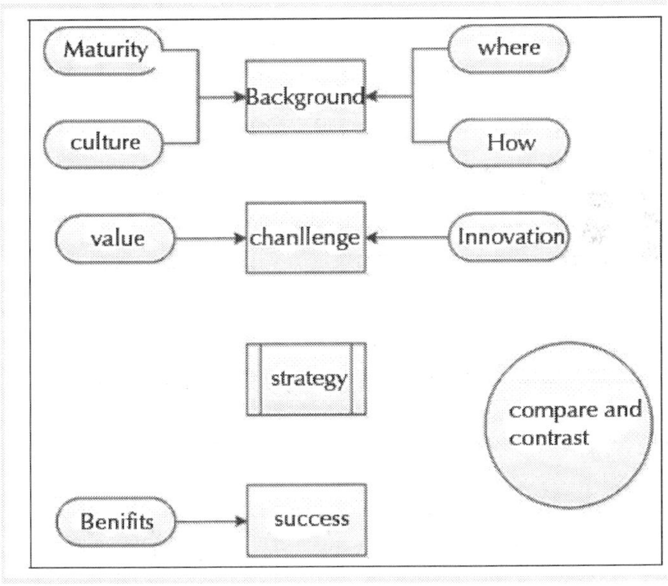


**Figure 1:** Structure of case studies.

Firstly, some background information on the projects should be presented. Since how and where BIM is adopted depends on the problems of specific projects, differences of background should be identified. The effect of different backgrounds and maturities of BIM on problem and solutions should also be analysed. Secondly, the challenges of each project should be discussed to identify why the case is worth being analysed. The research gap regarding BIM in this area should be identified. The main problem should be illustrated as well as the disadvantages of adopting traditional methods. This process will reveal the value and the innovation of the project. The third part of case study will present the strategy adopted in the projects to overcome the challenges. For the purpose of evaluation, the following questions should be answered: "What is the difference between this project and others?" "What is the main problem of the target case?" "What is the conventional solution?" "What is the reason for adopting BIM?" "Is this approach a good solution for the problem?" For the purpose of comparing and contrasting, the following questions, among others, should be answered: "For the same problem, what would the solution be on a different background?" "Is there any effect of the background on the difference of approach?" Finally, the success of otherwise of the project will be discussed. Any benefits provided by BIM should be discussed. Questions to ask include: "Does the solution improve productivity compared to conventional approaches?" "If the solution was adopted in another project, would the productivity be improved or not?". After this four-step analysis, the unique background of each project will be identified. Challenges, aims, strategies, benefits of different projects will be compared. A conclusion will be made as to which phase of the building lifecycle benefits most from BIM and which phase has the most potential to adopt BIM in different districts. The maturity of BIM adoption in the AEC industry in different countries will also be discussed.

## CASE STUDIES

As stated above, five projects will be described in this section focusing on different phases throughout the building lifecycle – design, construction and operation. BIM could be adopted in all these three phases to enable easy sharing, integration and visualization of information to improve productivity [7].

# Chong Qing International Circus City

## *Background of project*

Large-scale public projects account for a large proportion of China's infrastructure. Some projects are related to people's livelihood. In recent years, China's fixed assets investment scale has reached RMB 10 trillion, of which 65% rely on basic construction. With the implementation of the proactive fiscal policy and other policies to expand domestic demand, investment in public works projects has grown rapidly. Therefore, it needs to be facilitated with advanced technology and scientific management. However, because of the particularity of large-scale public projects, there are many difficulties in management. (1) These projects involve a large amount of investment and long cycles, which indicates late reclaim of investment and high risk. (2) Collaboration needs take place across many fields. This makes for potential management risk may affect the schedule and quality of projects. (3) They all have a great influence on the society. If there is an accident, it will have a large negative impact on the general public. (4) They share the same aim of offering comprehensive benefits for society. Chongqing International Circus City is located in Chongqing Danzishi town, partition A. The total land area is 3.333 hectares with a total construction area of 37,200 square metres. Building features include the main performance hall, ancillary services and facilities, the domestication of animal space and office apartments. The Circus City is one of the top ten cultural buildings in Chongqing. The main performance hall construction area is 21,847 square metres with a building height of 49.78 metres and a seating capacity of 1,489. The architectural form of the main performance hall generates two mutually wrapped surfaces through two sets of curved spaces which enclose the circus theatre space. The inside surface is round and full, gently stretching the outer surface with a natural extension of the roof ribbon integrated with the ticket booth and the catering, retail and other commercial facilities.

## Challenge – Complex Model

In the theatre building, the lower part of the auditorium, the lobby, stage and other parts form a complex-shaped space. Thus, the conventional 2D design consists of many curved sections to show the details of each part precisely [8]. Since positioning in 2D drawings is difficult and lacks precision [9], it is hard to create an accurate as-built plan for on-site construction workers. BIM can improve the precision of the design and construction of such complex spaces with 3D models [10]. Analysis tools such as SAP2000 and CATIA should be adopted together with Revit Architecture. For example, as shown in Figure 2, the outside curtain wall of the main performance hall is a complicated double-curve surface, which is hard to model. The designer of this building needs to integrate the parameterized model of the curtain wall in CATIA with the room model in Autodesk Revit to generate a complete building information model.

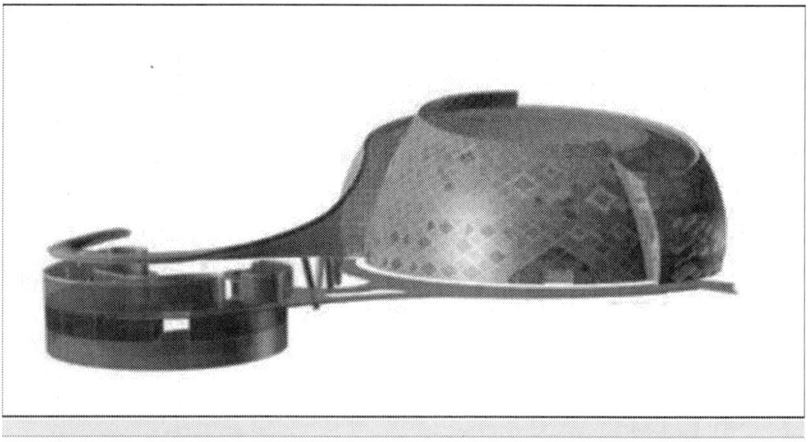

**Figure 2:** Outside curtain wall of main performance hall [17]

## Strategy

BIM is treated as the one of the standard streamline processes to generate drawings in the construction phase. It has been proven that visualization tools can enhance design comprehension and support collaborative work [11]. Massive software companies are providing new concepts for BIM solutions. Autodesk Revit, CATIA, AutoCAD, SAP2000, Midas and Fluent are chosen as the design and analysis tools in this case study.

The adoption of BIM includes the following five processes:

1. Create the building structure, facility and MEP 3D model based on Autodesk Revit software;
2. Integrate the parameterized model of the curtain wall in CATIA with the room model in Autodesk Revit to generate a complete building information model;
3. Conduct the structural analysis in SAP2000 and Midas software;
4. Conduct the calculation of material consumption and budget of building structure, facility and MEP qualitatively and quantificational based on the 3D information model aided design;
5. Conduct corporation with sub-contractors based on the 3D model.

In the initial design phase, CAD drawings are imported into Autodesk Revit software to create the BIM model of Chongqing Circus City. It includes a structural model, facility model (HVAC, water supply and drainage and fire extinguisher systems), electrical model and site model. Through the function "copy and monitoring" in Autodesk Revit, real-time information updates can be conducted in an inter-model way. Through the creation of 3D models, designers can understand the complicated construction space clearly, thus avoiding the disadvantages of 2D design. With 3D models, construction work can be guided precisely.

1. Optimization and visualization of outside curtain wall

The outside curtain wall of the main performance hall is a complicated double-curve surface. In the deep level of the optimization plan, CATIA software is adopted to create the model of the outside curtain wall, including the metal sheets, keel (Figure 3), ring beam structure, the curved structural column (Figure 4) and other components relevant to the curved surface of the curtain wall. Rhinoceros software is adopted to integrate all those components into Autodesk Revit.

2. BIM-model-aided complicated space design

As explained above, the lower part of auditorium, the lobby, stage and other parts form a complex-shaped space. The availability of this space needs analysis with the help of the three-dimensional modelling. The Circus performance hall is separated by a curved wall, which is both the structural element and modeling element.

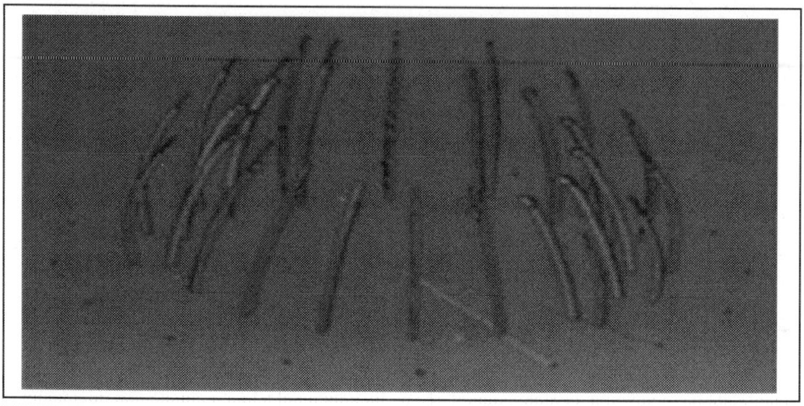

**Figure 3:** Main structural column [18].

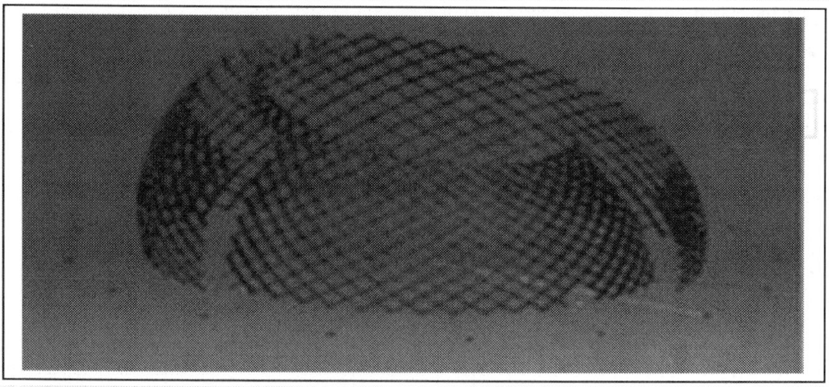

**Figure 4:** Keel of curtain wall [19].

The curved wall positioning constraints are subject to a number of conditions. Firstly, the curved wall in the first floor needs to ensure the reasonable width of cloakroom and shop. Secondly, the curve line of the curved wall on the second floor needs to guarantee an effective width and height of the audience lounge to avoid pressure. Last but not least, the intersection of the curtain wall and the curve wall should be below the structure of the roof. Based on the functionality of generating multiple planes and bisects, designers could achieve a satisfactory location of the curve planes to achieve the final space effect of the front hall. Structure, equipment, electrical and other sections need to be imported into professional analysis software for qualitative analysis and quantitative calculation of the building programme. A BIM model is interoperable and comes in different formats. Integrated information in the model could be exported to other software. This reduces repetitive tasks in the traditional design process, which improves the efficiency of the work of the designer. With the import of an AutoCAD Autodesk Revit model into Fluent software, architectural numerical simulation of wind loads are used to determine the actual wind pressure distribution to provide an accurate basis for the structure, curtain wall and structural design. In SAP2000 and Midas, structure calculation and structural deformation analysis of the construction of the main structure are analysed. Models are corrected in Autodesk Revit based on the calculation results.

3. Composite design and clash detection of line pipe

The aim of composite design and clash detection of line pipe is to avoid clash and confliction. Conventional line pipe in 2D design draws key sectioning planes to coordinate each section. However, in 3D design, BIM could be adapted to reflect the conditions in each section. Autodesk software is adopted to conduct clash detection in order to improve the productivity of design and clash detection of line pipe. The performance hall of Circus Santo's electromechanical space is mainly related to the mechanical and electrical line in the basement of the two story underground ventilation. It includes air conditioning systems, air-conditioning water systems, drainage systems, fire safety systems and electrical systems. 3D model clash detection will produce a large number of collision errors. In order to make clash checks clear, separate clash detection is conducted between civil structures, mechanical and electrical systems and electromechanical systems.

## Benefits of Adopting BIM

BIM is adopted in this project to conduct complicated modelling for construction site. Compared to conventional 2D design drawing, precision of positioning is improved for design and on-site construction work. The complexity of this model makes it very difficult to build the model with normal BIM design tools only, such as Revit architecture. BIM analysis tools are also adopted to facilitate the complex modelling. This also enables collaboration work with other software to create models more accurately. The process of clash detection is also simplified through pre-handling with BIM software.

## Wuhan New City International Expo Centre

### *Background and Challenge*

Wuhan New City International Expo Centre is planned to cover a total area of 6253 acres and a net area of 4270 acres. The central area of the planned construction area is 1.2 million square metres. An overview of the as-built plan for this project is shown in Figure 5. The exhibition hall of the first term has been completed for the final phase. Similarly to the last case, Wuhan New City International Expo Centre is a large-scale public project. There are traditional investment, duration and coordination problems in the project. Additionally, there are issues of public safety and a requirement for comprehensive benefits. The following characteristics of BIM could help to solve these issues: (1) BIM could simulate the construction phase to find potential issues; (2) a BIM model could facilitate collaboration with a visualized platform and central database; (3) BIM could integrate the facility management information with the 3D model to guide the asset management and transportation.

**Figure 5:** As-built plan of Wuhan New City International Expo Centre.

BIM applications in this case are mainly focused on operation and maintenance. According to the actual progress of the International Expo Centre and the digital expo service requirements, the adoption of a BIM model will focus on investment, exhibition, traffic signs, guidance, equipment and other aspects of emergency services.

## Strategy

According to the BIM regulation of planning [12], the BIM construction plan of Wuhan New City International Expo Centre needs to be created first. The BIM model is created based on the construction plan (Figure 6). The properties are embedded into the BIM model based on the operational purpose.

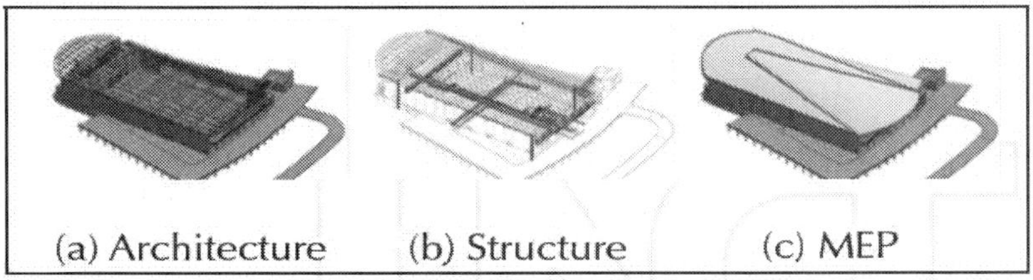


**Figure 6:** BIM model of Wuhan New City International Expo Centre.

1. Analysis of traffic organization

Navigation systems are embedded into the integrated model, which includes information on category, ID, location of installation, materials, technology, text messages, font size and position, indoor logos, outdoor logos and car-park logos. BIM also enables checking of the clarity of view at the entrance and exit. The usability of signs for guidance could be observed in the simulated scenario. For example, as shown in Figure 7, the walkthrough in a third-person viewpoint experience proves that the glass protective structure located in the broad platform of the second floor plays a very important role in guiding people's navigation.

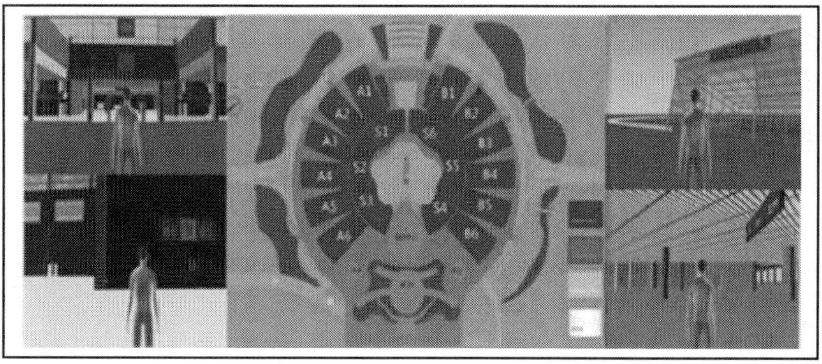

**Figure 7:** Using BIM for guiding transportation

2. Space management

In the operation phase, the main purpose of the pavilion is for exhibition. Thus the space management includes both the conventional space for MEP equipment and the partition of the pavilion's interior space. The properties and status of space change over time. For ease of management, properties relevant to the exhibition are assigned for different partitions. These properties include the location of the partition, status of lease, lease holder, etc. (Figure 8)

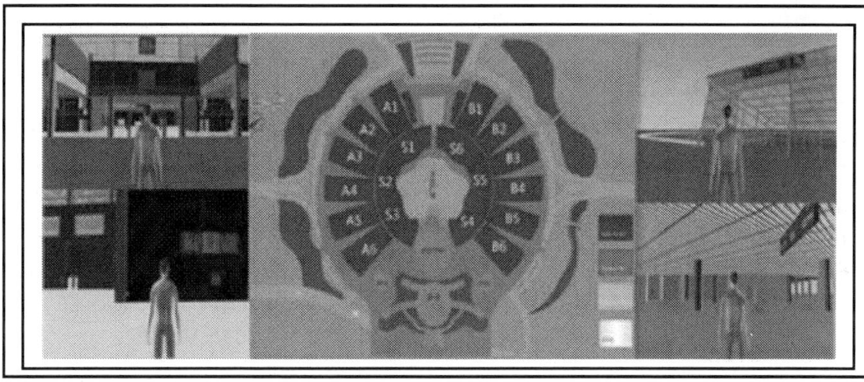

**Figure 8:** Exhibition partition properties.

### *Benefits of Adopting BIM*

The BIM model of Wuhan New City International Expo Centre facilitates the rescue of equipment in emergencies, investment for exhibitions, and access to information via a central database. An operational management module based on this model provides support for emergencies and digital expos. Operational safety and security are improved and the public are able to learn about the project. Thus the overall value of the project is improved.

## Chengdu Financial City

### *Background of Project*

The Chengdu Financial City is located in the City of Chengdu, on TianFu Avenue. It covers 170,000 square metres with a height of 210 metres, including offices in the South Tower, dormitories in the North Tower and a basement level. The core-tube system is applied in the South Tower and the frame core-tube system in North Tower. The total investment for this project was RMB 1.3 billion. The financial city consists of trading market, public plaza, shopping mall, office buildings, living quarters, etc. This project includes three sections: architecture construction model, structural model and Mechanical, Electrical and Plumbing

(MEP) model. Navisworks is adopted to conduct clash detection and visual collaboration for the large integrated BIM model. Please refer to Figure 9 for the adoption of BIM in the construction phase. As the financial city is used comprehensively for different purposes, it is important to ensure every part will be occupied. Otherwise, vacancy will consume a huge amount of funds. The adoption of BIM could enable the user to view the as-built plan and the function of each part. The visualization function of BIM can attract potential contractors in the design stage as well. Merchants such as Wal-Mart have their own regulations for space. Adopting BIM in the design stage can thus also enable the owner to arrange the functionality of different parts corresponding to their different rules.

## *Challenge*

In order to ensure every part will be occupied after the construction phase, the whole BIM model of this massive financial city must be built in the design phase. Arrangement of different functionalities for each part should be completed in this phase as well. It is a great challenge to adopt BIM throughout the construction phase, including creating an infrastructure model and MEP model and a clash detection and construction plan. Massive problems may occur when keeping all the data classified efficiently and synchronized in real time.

## *Strategy*

1 Creation of infrastructure model

This part is mainly focused on the structural model for checking collisions of MEP. It consists of foundation, raft foundation, retaining wall, concrete beam, steel column structure, slab, shear wall, partition wall, curtain wall, stair, door and windows. The structural BIM models are all referred to the type of steel column, including flange, web and R angle. If there are any changes, the type of steel column can be changed efficiently. Drawings from different professions can be updated synchronously to avoid nonsynchronous problems.

2. Creation of MEP model

The MEP model includes power distribution, telecommunication, ventilation of air conditioner, water supply, protection of fire and water gutter system. Different tasks are assigned to engineers in different areas in a collaborated way. They can work on the same model synchronously. Engineers check the model for collisions between pipelines and mark up the model for discussion in the next meeting. To execute editing and inspection tasks in BIM software, a colour scheme is distributed to each type of MEP component. For example, red is assigned to power distribution components; yellow is assigned to water supply and green is assigned to fire protection.

3. Clash detection

Conventional clash detection in the synthesis process of pipelines is conducted by professional engineers manually. It is a massive and time-consuming job. Carelessness and mistakes always occur in manual work. The result is wastage of both time and budget. Thus, plans need to be revised in real time on site. Through clash detection in BIM software, hard collisions can be identified visually and automatically. The time needed to change plans can be potentially reduced and the value of space could be improved by optimization. In this project, a 30% Return on Investment (ROI) is achieved through BIM.

4. Construction plan

The construction plan can be simulated in 4D BIM software. For example, simulation of the working route of a crane can be simulated visually. The plan can be simulated and visually displayed on the projector for collaboration and discussion, in a highlight colour for easy recognition. The schedule can be updated with the BIM model. There are even 5D simulations which include the budget based on materials. The client is developing a standard database for the calculation of total cost of materials.

5. BIM model for facility management in operational phase

After the construction phase, an as-built BIM model is adopted for facility management. This model serves as a central database. The platform needs to be developed for different areas. Basic facility management is split into three sections: maintenance plan, assessment management and space management. The advantage of adopting BIM to conduct facility management is the 3D environment, which visually displays all the objects clearly. All of the maintenance documents can be updated in real time. Relevant manuals can be searched efficiently and specific components can be located effectively. BIM is adopted throughout the construction phase of this project. Most parts of the structure and pipeline are digitalized. Pre-handling of clash detection with integrated rules reduces costs and time. The as-built model can also be treated as a digital asset of the building, which can be used throughout the operational phase.

### Benefits of Adopting BIM

BIM is adopted in this project to design the whole BIM model corresponding to different regulations of merchants and to arrange various functions for each part. Most of the financial city will be occupied as soon as it is ready to avoid vacancy and waste of funds.

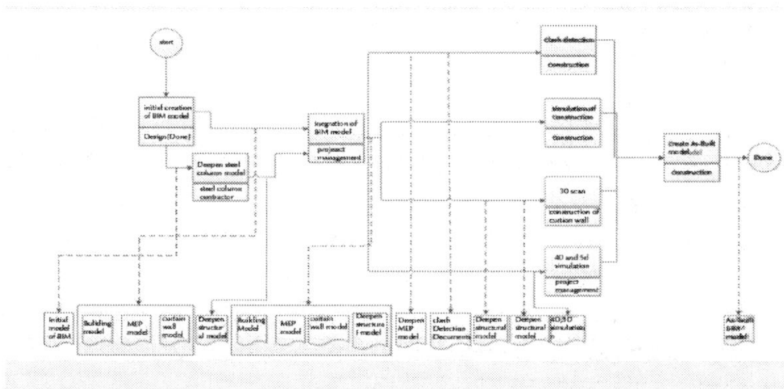


**Figure 9:** The adoption of BIM in the construction phase.

## Shanghai Disaster Control Centre

### Background and Challenge

Shanghai Disaster Control Centre is one of the most important information systems of the State Grid Corporation. It consists of five floors covering 28,124 square metres, comprising one underground level and four above ground. Total investment in this project is RMB 300 million. The basement level will be occupied by early warning equipment. The ground floors comprise of computer rooms, air-conditioning rooms and a command centre for emergencies. The early warning equipment is of vital importance and needs to run all the time. It is critical to monitor the status of all the equipment in order to provide urgent support quickly.

## Strategy

Traditionally, sensor and warning systems are embedded into disaster control centres [13]. When an emergency happens, the facility manager will be warned in the command centre. However, it is difficult to identify which type of disaster is occurring and in which facility. It is difficult to recognize the real-time status of a facility. A BIM facility management system can achieve visualization and real-time monitoring of all facilities with communication to sensors embedded in facilities. Wang and Chen developed an approach to conducting tele-inspection based on virtual space [14]. Operation management involves not only facilities management, but also management of space, traffic and so on. The establishment of the operation management platform will be a complex and arduous task. To avoid development risks and shorten the development cycle, the development of a BIM operation management platform only focuses on the centralized management of the facilities, including the air-conditioning and ventilation systems, water-chilling systems, water-cooling systems, water-supply and drainage systems and electrical power systems. This project aims to develop an integrated platform to manage the operational information of this building, in two ways: equipment information monitoring and dynamic data display and warning [15].

1. Equipment information monitoring

Equipment information monitoring mainly refers to the monitoring and control system in the context of the whole building, including a regional view, a sub-system view, a system information view and a facility information view, and the corresponding auxiliary functions. The BIM operation management platform in the Shanghai Disaster Control Centre is divided into five regions according to the floors: B1 subzone, 1f subzone, 2f subzone, 3f subzone, and 4fp subzone. Zones are defined on each floor according to the function of the rooms. B1 includes five areas: cooling tower water inlet and air-conditioning rooms, fire-control pump rooms and life pump rooms, refrigerator rooms, UPS and battery rooms, and distribution rooms. The Centre of

Disaster Recovery is thus divided into 17 partitions in the BIM operation management platform. In the process of BIM model processing, each partition is highlighted. When any area is selected, the main interface of the BIM operation management system shows all the system in the area. From each regional view, it is shown clearly whether the partition within the system and the equipment is working properly. Through the development of auxiliary functions such as translation, zooming, and roaming, the user can find any devices in the partition efficiently and conveniently. Then the user can read the physical parameters and real-time operation parameters to understand the working state of the equipment. The system interface is shown in Figure 10.

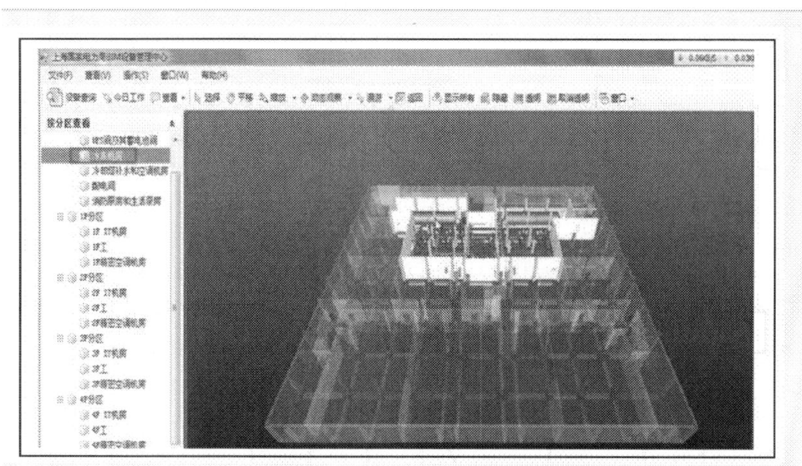

**Figure 10:** System interface of the BIM-enabled FM system.

The subsystem view mainly divides up zones according to the MEP system for the convenient supervision of the system's running status. As in Figure 10, the MEP system in the BIM model is divided into 15 big systems: water supply and drainage system, VRV air-conditioning system, mechanical smoke-extraction system, mechanical pressed-air supply system, emergency exhaust system, gas-extinguishing exhaust system, normal exhaust and gas-extinguishing exhaust system, air-conditioning and fresh-air system, K(X)-1 fresh-air system, TEF-RF-1 system, basement fresh-air system, basement fresh air system without processing, basement ventilation system, water cooling system, and water freezing system.

2. Dynamic data display and warning

The dynamic representation of data and alarms include searches of the facilities, dynamic location, display of the facilities and the maintenance routes, relative records of the daily maintenance work and the alarm information of the facilities. The facility manager can search and check the facilities by type, number, name and time of purchase. People can also check the attributes of the facilities, the maintenance routes and records, and previous data directly. When choosing any facility, when clicking the viewing button, a user can find the relevant facility in the BIM model. Attributes and the current time operation situations of the facility are listed in the attribution column. Figure 11 shows the 3D model and dynamic data information of the previously checked switch cubicle. The images imitates real eyesight from the spatial location. The checked switch cubicle is highlighted as differing from other facilities. The right attribute column shows the relative information, such as the number, the name, the type, the current floor and room, the date bought and last maintained, the maintenance cycle, the information of the switch, the line, the actual three-phase current, the threephase voltage values, the P value and Q value, etc. With this information, the worker can judge the operation state of the switch cubicle rapidly and directly. Meanwhile, they can find the accurate location, check the maintenance records and check whether the part needs maintenance.

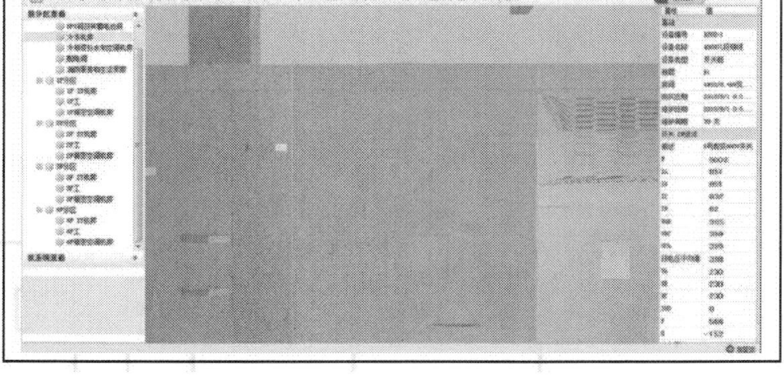

**Figure 11:** 3D model and dynamic data information of the previously checked switch cubicle.

If any facility is chosen, the system will match the corresponding maintenance routes in red. These routes are typed into the backstage database in advance. Workers can refer to or print the routes to find the facilities' location, which is very convenient. Another important function is facility alarm. When facilities break down, the BIM operation management platform will get a notification and obtain details about the hardware and the alarm in good time. On the alarm interface, workers can find basic information such as the current floor and the current room, which will help workers to find the facilities quickly.

### *Benefits of Adopting BIM*

Adoption of BIM enables the real-time monitoring of facility status. Thus, the user will discover issues quickly. Positioning and navigation could also be realized in a BIM database. With the assistance of BIM, a facility manager can get to the right spot quickly. Time is saved and additional losses avoided. A BIM operation management platform will also automatically connect the maintenance database and come up with the relevant plans.

## National Synchrotron Radiation Research Centre (NSRRC)

### *Background and Challenge*

The Taiwan Photon Source (TPS) project aims to create the most advanced third-generation synchrotron acceleration facility in the world. The budget is TWD 6.885 billion, including RMB 2.78 billion for civil engineering and MEP and RMB 4.105 billion for the acceleration facility. Figure 12 shows the as-built model of this project. There will be another TWD 8 billion for the experiments in this research centre. As shown in Figure 13, the buildings constitute an administrated lapping building, the YiGuang Building, the Taiwan Light Source, an electromechanical building, a research centre and a TPS-administrated operation centre.

**Figure 12:** The as-built overall view of TPC [20].

A large number of valuable facilities are found in this centre. Similarly to the Chengdu Financial City, this project requires real-time positioning and a high level of safety in order to keep all facilities in good status. The difference between them is that this project is focused more on maintenance and space allocation. Since all of the facilities are precise and valuable, working conditions are very strict. For example, small issues such as leaks may damage the facility and cause big losses. Thus, maintenance of the working environment such as waterproofing materials for building should be changed regularly. Manual handling of the maintenance schedule is imprecise and unreliable [16]. Space allocation is also critical in order to avoid inappropriate design of space or negative effects on researchers' sense of well-being. Space allocation will be designed with the help of the researchers and engineers themselves. Thus, a collaborative platform is needed to visualize the as-built plan.

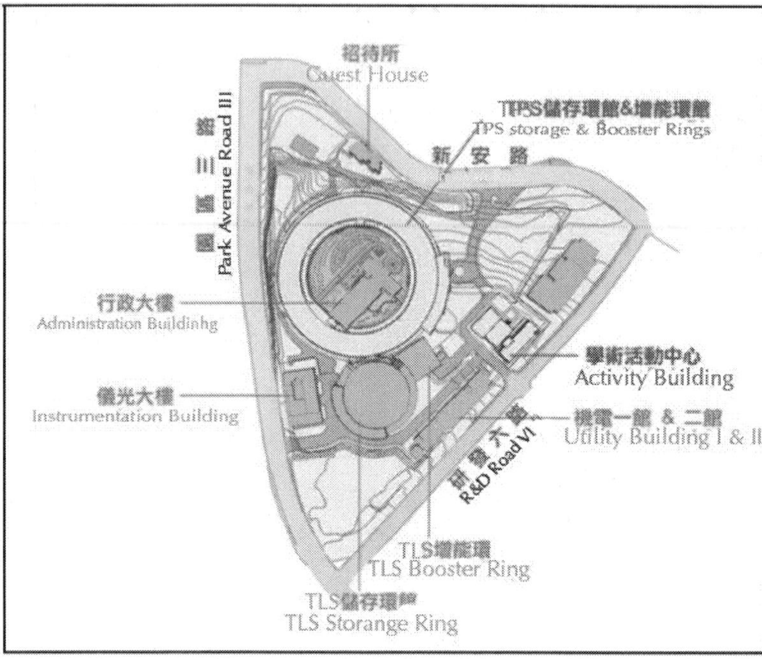


**Figure 13:** Functions of different zones of TPS [21].

## *Strategy*

1. BIM-enabled maintenance schedule and warning

The lifecycle of buildings could be as long as 40 to 50 years. However, materials in buildings cannot survive for that long. Maintenance schedules for different materials vary. The facility manager cannot remember all the lifecycles of different materials. Conventionally, maintenance is conducted after issues arise. Facility managers may also be changed during the lifecycle of a building, resulting in incomplete delivery of maintenance documents. BIM integrates all the maintenance documents in a 3D model, linking each material, component and facility with maintenance schedules, properties, manufacturers and other relevant information. This project adopted BIM to ensure early warnings about scheduled maintenance and provide reminders about materials whose lifecycles were ending. Cost information was linked to different types of materials for budget calculation. Detailed information on manufacturers, specifications, etc., is attached to objects as properties. These data are used to arrange work and impose order on materials of the same type. For ease of access to different information, a search engine is developed to retrieve information by name of building, name of facility, code of facility, etc. Once a part of the building requires maintenance because a cable is damaged or materials expire, detailed information could be identified efficiently from this engine. A user will be guided to the specific position in the BIM model. 2) Space allocation and modification Light sources in a synchronous acceleration facility are indispensable in basic scientific research. For a limited budget, space designers are not employed for all laboratories. To avoid inappropriate design of space or negative effects on researchers' well-being, space allocation will be designed by the researchers and engineers themselves. Without the adoption of BIM, this allocation must be conducted with 2D drawings. However, most of the researchers are not from a civil engineering background. Due to the lack of sense of space, BIM is thus adopted to provide an immersive feeling. They can then mark into the model an appropriate working space before the integration of a complete plan of space allocation. 3) Asset management The NSRRC project involves massive research facilities. It is hard to identify whether any part of these facilities is damaged. However, with the technology Radio Frequency Identification (RFID) and BIM, a unique identification can be assigned to all the facilities. They can be tracked remotely in the system. If they do not remain in the initial place, a warning will be sent to the manager automatically. A 2D barcode system and mobile devices such as mobile phones and PDAs could be integrated into this system for search references of all facilities.

## Benefits of adopting BIM

Adoption of BIM in this project enables collaboration in the pre-design stage taking account of researchers' own demands for space allocation. BIM is also adopted in this project to monitor the real-time status of facilities, similarly to the Shanghai Disaster Control Centre. Unlike the last project, BIM is applied to manage the maintenance schedule more efficiently with automatic notification. This ensures a good working environment for all the facilities and avoids ageing materials causing damage to facilities.

# CONCLUSION

This paper has presented five cases of BIM adoption in the design, construction and operational phases under different circumstances. Methodology is developed to analyse, compare and contrast BIM use in these five projects. By comparing the five construction projects in Asia, it is concluded that in the design phase, BIM is mostly adopted to conduct complicated modelling for construction sites and to enable non-constructors to change plans based on their preferences. The analysis function of BIM software facilitates the improvement of the precision of drawings. In the construction phase, positioning will be more precise with the adoption of BIM. Building plans can be simulated and optimized to improve safety and productivity. Drawings from different professions are imported into BIM software to conduct clash detection. Clash detection is implemented in BIM simulation software in order to check if there is any collision in the MEP model. In the operational phase, BIM helps to monitor the equipment in real time and ensure it works normally to avoid losses caused by disaster. Digital expos and public emergency support are assisted by BIM through improved operational management. Moreover, the as-built model can be treated as a digital asset of the building, which could be used throughout the building life cycle.

# ACKNOWLEDGEMENTS

The authors would like to thank the Sichuan Southwest Project Management & Consultancy Co. Limited Company and China Construction Design International (CCDI) Company for providing project data. This work is also partially sponsored by Shanghai International Enterprise Cooperation Program of STCSM (Science and Technology Committee of Shanghai Municipality) under Grant number of 12510701700.

# REFERENCES

1. Wang, X. and Dunston, P.S. (2006). "Compatibility Issues in Augmented Reality Systems for AEC: An Experimental Prototype Study". Automation in Construction, Elsevier, 15 (3), 314-326.

2. Schlueter, A. and Thesseling, F. (2009). "Building information model based energy/exergy performance assessment in early design stages". Automation in Construction, 18, 153-163.
3. Bormann, A. and Rank, E. (2009). "Specification and implementation of directional operators in a 3D spatial query language for building information models". Advanced Engineering Informatics, 23, 32-44.
4. Wang, X., and Love, P. (2012). BIM+ AR: Onsite information sharing and communication via advanced visualization. Computer Supported Cooperative Work in Design (CSCWD), 2012 IEEE 16th International Conference, 850-855.
5. Wang, X. (2007). "Using Augmented Reality to Plan Virtual Construction Worksite." International Journal of Advanced Robotic Systems, 4 (4), 501-512s
6. Hardin, B. (2011). BIM and construction management: proven tools, methods, and workflows, Wiley Publishing, IN, Indianapolis.
7. Singh, V., Gu, N. & Wang, X. (2011). A theoretical framework of a BIM-based multi-disciplinary collaboration platform. Automation in Construction, 20, 134-144.
8. Syal, M., Parfitt, M. K. and Willenbrock, J. (1991). Computer-integrated Design Drawing, Cost Estimating, and Construction Scheduling. Pennsylvania State University, Department of Civil Engineering, NAHB/NRC Designated Housing Research Centre
9. How, H. (2002). Precision data acquisition using magnetomechanical transducer. Google Patents.
10. Wang, X. (2012). "Extending Building Information Modelling (BIM): A Review of the BIM Handbook." Australasian Journal of Construction Economics and Building, 12 (3), 101-102.
11. Dunston, P.S. and Wang, X. (2005). "Mixed Realitybased Visualization Interfaces for the AEC Industry." Journal of Construction Engineering and Management, American Society of Civil Engineers (ASCE), 131(12), 1301-1309.
12. Eastman, C., Teicholz, P., Sacks, R., and Liston, K. (2011). BIM handbook: A guide to building information modeling for owners, managers, designers, engineers and contractors. Wiley. Hoboken (New Jersey).

13. Igarashi, A. and Kobayashi, T. (1990). Disaster prevention monitoring and control facility. Google Patents.
14. Wang, X. and Chen, R. (2008). "An Empirical Study on Augmented Virtuality Space for Tele-Inspection of Built Environments", Journal of Tsinghua Science and Technology (Engineering Index), Elsevier Science, ISSN: 1007-0214, 13 (S1), 286-291.
15. Wang, J., Li S., Wang X., Mao C., and Guo J. (2013). "The Application of BIM-Enabled Facility Management System in Complex Building." 10 Int J Adv Robot Syst, 2014, 11:124 | doi: 10.5772/58447International Journal of 3-D Information Modeling (IJ3DIM), 2(3), 16-33.
16. Akcamete, A., Akinci, B., and Garrett, J. (2010). Potential utilization of building information models for planning maintenance activities. Proceedings of the International Conference on Computing in Civil and Building Engineering, 151-157.
17. Zhulong BIM (2013). Outside curtain wall of main performance hall. [image online] Available at: http://bim.zhulong.com/case/read171253.html Accessed on 25 May 2013.
18. Zhulong BIM (2013). Main structural column. Available at: http://bim.zhulong.com/case/read171253.html Accessed on 25 May 2013

Chapter 4

# Performance Measurement in a Quality Management System

A.R. Rezaei, T. Çelik, Y. Baalousha

Department of Civil Engineering, Eastern Mediterranean University, Famagusta, Northern Cyprus, Via Mersin 10, Turkey

## ABSTRACT

Quality management system implementation has become a must for construction companies in some countries to be able to enter tenders. One of the most common quality standards is the ISO 9000 quality management standard and many companies seek ISO 9000 certification in today's highly competitive market. However, in getting this certification, most companies face difficulties, such as the high amount of paperwork, improper documentation, poor communication among employees and project participants, and low employee morale as a result of lack of motivation. In this study, a web-based office automation system was developed. Web facilities and the database management capabilities of Microsoft Visual Studio 2008 were applied to create a data warehouse that was aimed to reduce paperwork, create a proper documentation system, improve communication, and calculate employee performances, in order to create a motivation system for company personnel. Short-term feedback of the practical implementation of the system demonstrated its practicality and advantages, and the positive view of the managers. Also, it is anticipated that long-term feedbacks would also prove its appropriateness and ease of use.

# INTRODUCTION

Quality Management System (QMS) certification has become a must in today's highly competitive construction market. One of the most common methods of quality management system application to a construction company is ISO 9000 certification. While the reason for many companies wishing to achieve a quality management certificate, like ISO 9000, is only for eligibility to enter tenders, some other companies seek for ISO certification to benefit genuinely from its numerous advantages. However, construction companies face a number of difficulties in the certification process, including an increase in paperwork, an improper documentation system and poor communication among personnel. As a result, these may cause re-working, as mentioned by Alshawi and Ingirige [1], low worker interest in applying new working methods [2], [3], [4] and [5], and low employee morale and motivation. These problems make the certification process arduous and thus some companies might even give up.

The aforementioned problems will, on the other hand, influence company performance. Literature shows a considerable increase in the field of supply chain performance and its effects on company performance [6],[7], [8], [9], [10], [11], [12], [13], [14], [15], [16], [17], [18] and [19], and company performance management and measurement [20], [21], [22], [23], [24] and [25]. The majority of researchers used structural equation modeling and, through the development of a number of hypotheses, tried to prove or reject them based on a survey among different experts. However, research on performance quantification is relatively scarce, and less attention has been paid to employee performance.

A number of office automation systems are currently being used, such as Autodesk Constructware, Business Collaborator, and e.Build.ca. Autodesk Constructware creates a centralized database of all project documents, which can be accessed by all project participants, and which is capable of managing documents, design, bidding, costs, construction and operations. Business Collaborator facilitates collaboration across teams, automates processes and enables access to project information. It is only aimed to automate traditional paper-based processes to improve response times, reduce errors and cut costs. The Collaborative Business Platform from e.Build.ca is an online collaboration and project management tool designed for the planning, design and implementation of construction projects. However, these programs were not designed for measuring employee performance, or provide a decision tool for managers to monitor employees, develop an Organizational Breakdown Structure (OBS) with appropriate lines of communications for ease and inform recipients by messages on his mobile phone to save time and enhance performance.

This research fills this gap by developing a system to measure the performance of personnel and subsequently apply the obtained performance factor to motivate and appreciate the personnel, which in-turn affects employee performance and, consequently, company performance.

The objectives of the present research study were to overcome QMS deployment problems and create a decision tool for company managers to track employee performances. The data warehousing technique and soft computing methods were applied to develop a system called Performance Management Support System (PMSS). PMSS is a web-based office automation system that measures the performance of employees based on three management factors of time, cost and quality. Each of these indexes could provide useful information for managers [26] and are emphasized by many researchers [27], [28] and [29]. The measured performances can further affect their standing in the company and also their salaries. This is highly effective on the motivation of personnel, as a result, creating a competitive environment among personnel and encouraging them to undertake their responsibilities in the shortest time with the lowest costs and at the highest quality level. Thus it overcomes QMS deployment difficulties and ultimately imposes continuous improvement. PMSS is generally designed and developed for every organization type; however, due to the more complex nature and uniqueness of construction projects and construction companies [30], the practicality of the developed system was checked through evaluations with experienced general managers of some construction companies and ISO certification agencies. Time saving, cost reduction, less paperwork, easier communication and more efficient documentation were affirmed to be the most important advantages of PMSS. The feedbacks approved the appropriateness and applicability of the system in the construction sector.

# LITERATURE REVIEW

## Quality Management

Quality management is a critical component in the successful management of construction projects. ISO 9000 certification is the most successful quality management system for many construction companies. However many researchers (e.g., [2], [3], [4], [5], [31], [32], [33], [34], [35], [36] and [37]) emphasized the difficulties of the certification process. These include an increase in paperwork, an improper documentation system and poor communication among personnel. Soft computing and Internet technology is believed to be a good solution for such problems [38] and [39].

Many attempts have been devoted to solving such problems and inconsistencies in the construction industry. Lam and Ng [2] used an Internet-facilitated quality management environment that aims to examine the potential of applying the web-based techniques to collect, manage, access and distribute quality-related data at the construction project level. Kim et al. [35] developed a Personal Digital Assistant (PDA) which used a wireless web-integrated system, and which was intended to perform quality inspection and the defect management of apartment housing projects. The ISO 9000 Quality Management

Information System (QMIS), developed by Chin et al. [32], was proposed to integrate scheduling with the quality system process. However, lack of enthusiasm still exists among workers and employees. They believe that the certification process and application of a quality standard will control their performance, and will act as a controller limiting them [40].

Thus, an Integrated Information System can help overcome quality management system deployment difficulties in a construction company.

# PERFORMANCE MANAGEMENT

In today's highly competitive market, traditional organizational management methods cannot be considered appropriate strategies. Recent developments in database management systems, Internet technology and office automation systems encourage managers to apply these methods to their company in order to survive and compete with their competitors. Performance measurement is one of the most important decision tools for managers.

A performance measurement system can be defined as a set of metrics used to quantify both the efficiency and effectiveness of actions [41] and [42]. Performance measurement methods are attractive to researchers, as stated by Phusavat et al. [43] and [44]. ISO 9001:2008 clearly specifies performance measurement as part of its requirement no. 8. Performance measurement helps to bring more scientific analysis into a decision-making process. It underlines the change towards management by information and knowledge, instead of primarily relying on experiences and judgment [43].

Since 1980s, the focus of performance measures shifted from purely financial factors to a combination of financial and non-financial ones. The factors affecting performance measurement in different research studies are based on one, or a combination of some criteria like finance, operations, quality, safety, personnel and customer satisfaction. Methods like the balanced scorecard, the performance pyramid, the performance measurement questionnaire, the results and determinants framework, the performance prism, the economic value added, and the Skandia navigator have recently been used [13], [45], [46], [47], [48],[49], [50], [51], [52], [53], [54], [55], [56] and [57].

Ho [21] states that performance refers to the achievements, in quality and quantity, of an individual or group work. Employees are critical components of business success [21] and their performances directly influence company performance [58]. Thus effective strategies to motivate and enhance employee competency are of urgent need for companies. Literature does not show much work on employee performance. Medlin and Green [59] examined the relationship between the constructs of employee engagement and employee optimism, as a means to improve employee performance. Kuo et al. [25] researched using productivity and attendance as evaluation criteria for employee performance. Tarng and Liu [60] developed a document management system that creates microfilm from all documents, which in turn could be used to evaluate employee performance.

Lack of proper links between information systems and performance measurement systems is highlighted in different studies [47] and [54]. They stated that advances in information technology, such as data warehousing and web-based technologies, can support performance measurement systems to achieve enhanced efficiency and effectiveness. Laitinen [61] emphasized the role of management information systems in delivering information to managers. He points out that a CEO spends, on average, 23% of total working time on getting information. He continues that, together with informing and reviewing information, information-handling occupies, on average, 51% of a total working day. The role of information systems and database systems on performance measurement was discussed in a study conducted by Phusavat et al. [43].

It can be concluded that the importance of company performance and the factors affecting it are stressed in different studies. However, a quantitative employee performance measurement method rarely exists. Thus it is believed that it is necessary to conduct a more comprehensive study on employee performance. Considering an Integrated Performance Measurement System (IPMS), a quantitative method using a data warehousing technique was, thus, developed, which calculates employee performance by using three classic project progress issues, such as cost, time and quality [62]. These metrics were emphasized in a number of research studies, to be the most important measurement factors of performance [63], [64], [65],[66] and [67].

Thus employee performance measurement and motivation of personnel could create a competitive environment among them, which in-turn would help managers to control their subordinates, together with enhancing company performance.

# DATA WAREHOUSING

Enterprises are increasingly turning to software systems to seek support for enterprise performance measures [68]. Companies need to have proper information systems and database management systems to capture and keep different information, which are their most valuable resources.

Successful support of managerial decision-making is critically dependent upon the availability of integrated, high quality information, organized and presented in a timely and easily understood manner [69]. Despite the growing need for more information, every day organizations create billions of bytes of data about all aspects of their business. It is estimated that only a small fraction of the data that are captured, processed and stored within the company is actually ever made available to executives and decision makers [70]. The concept of a Data Warehouse (DW) is part of the response by IT to meet this need, as pointed out by Golnabi et al. [71]. A functional DW organizes and stores all available data needed for informational and analytical processing over a historical time perspective [72], and provides decision makers with consistent, timely, reliable and accessible data without a negative impact on the operational systems from which the data is extracted [68] and [73].

DW by itself does not create value. The value comes from the use of its data in other applications. Another fact is that a DW is not a ready package to be bought and used by a company; instead, it should be specifically designed to meet company needs. This idea encourages companies to develop their own DW and appropriate computer applications to solve and overcome their difficulties, and take the maximum benefit out of data warehousing capabilities.

Jarke et al. [73] define a DW as a structured extensible environment designed for the analysis of non-volatile data, logically and physically transformed from multiple source applications to align with business structure terms, and summarized for quick analysis. An important concept of a DW is that its data comes from one or more operational applications, manipulated into a common format for the warehouse, and inserted into the warehouse with any necessary calculations or additional appended data. Then, the data are loaded into appropriate reference tables for efficient query performance, analysis, reporting or data mining by the user, through different available tools, such as web applications or crystal reports. A relevant data mining tool is On-Line Analytical Processing (OLAP), which provides a service for accessing, viewing and analyzing large volumes of data with high flexibility and performance. The essential characteristic of OLAP is that it enables a numerical and statistical analysis of data organized in a multi-dimension [72] and [74].

DW typically uses multidimensional, as well as relational, storage structures [68]. The multidimensional structure physically stores the data in array-like structures that are similar to a data cube. In the relational structure, the data are stored in a relational database using a special schema (star or snowflake) instead of a traditional design. The developed system in the present study uses a relational structure, considering star schema.

Multidimensional OLAP is used for a multidimensional database system, and a relational OLAP for a relational database system. The relational OLAP is designed to remove the technical constraints of traditional data access methods, resulting in a faster response and ease of use [75].

A number of research studies on DW applications in the construction industry [76], [77], [78], [79] and [80]show that it has considerable application potential. Historical data storage and access, data extraction from different internal and external data sources, data cleansing and analysis, and OLAP capability in report retrieving are main characteristics of data warehousing, which encouraged authors to consider it for the present system development.

Different departments exist in a construction company, and each of them has its own activities and work procedures. These departments may use their own Information Systems (IS). On the other hand, a company normally has several projects that could also have their own IS. These systems are considered as data marts for the main office data warehouse. A data mart is a subset of an organizational data store, usually oriented to a specific purpose. The developed system uses the data marts to calculate employee performances and provide different reports, using OLAP capabilities.

Thus, the data warehousing technique could be a proper solution for companies to develop integrated database management systems and decision tools.

# METHODOLOGY AND SYSTEM DEVELOPMENT

The methodology was based on developing a web-based office automation system for performance measurement using a data warehousing technique. Considering the results of a literature survey, the system development took place in almost two years. Then consulting with some experts in both fields of management and IT, the primary pilot model was revised and the present system was developed and ready to be practically checked at the end of 2009.

The developed system consists of three major parts including databases, user interfaces and web applications. MicrosoftVisual Studio 2008 was used to develop the main structure of the system. Microsoft Visual Studio is an Integrated Development Environment (IDE) from Microsoft, which can be used to develop console and graphical user interface applications along with web sites, web applications and web services [81]. The database of the system was created using the Microsoft SQL Server 2005, which is a Relational Database Management System (RDBMS). User interfaces and web pages of the system were designed and developed using ASP.NET which is used to build dynamic web sites, web applications and web services. C# (C Sharp) was used to connect database, user interfaces and web applications.

The OBS of the company will be developed by PMSS, and an employee will be assigned to each position in OBS. Personnel profiles and login accounts will then be created. The users can access the system through an authentication page. Different departments, projects, activities and their related data will then be defined. Users can use the message transfer mechanism of PMSS to send and receive work. The performance of each work undertaken is measured in three dimensions, namely time, cost and quality, and the performance of the employee will be calculated using the proposed formulas. A sample application diagram of PMSS is shown in Figure 1.

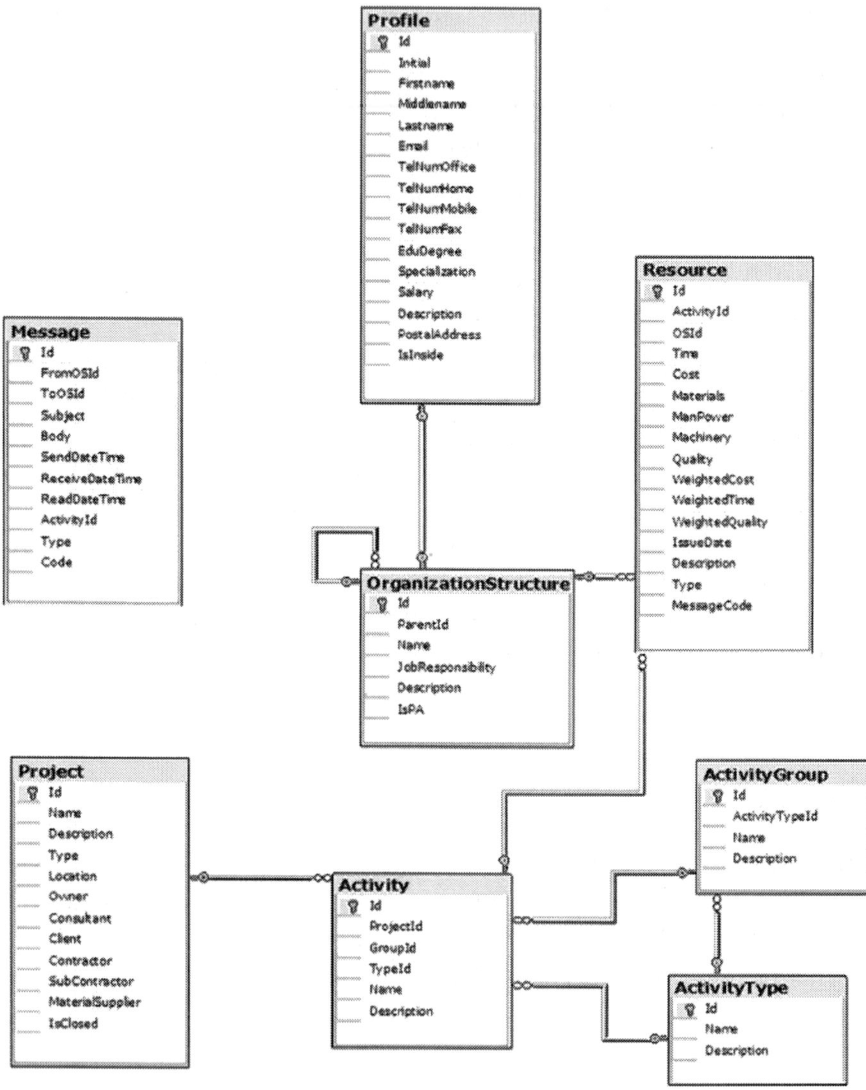


**Figure 1:** PMSS application diagram.

Various tables and databases are used to develop the system. An organization structure table is used to develop the OBS of the company.

The departments and profiles of company employees are defined in the profile table. Projects and their related data are entered into the system through the tables, named as project, activity type, activity group, activity and resource. The heart of the system is the message table, which has the capability of transferring data, sending and receiving messages, measuring performances and storing all reports. The DW then uses the databases from different departments and projects to extract the necessary data to perform the required queries. The message table, shown in Figure 1, is not physically connected to other tables, since it can also be used by external users like clients, consultants, contractors and material suppliers, which are not placed in the organizational structure of the company. Therefore, it uses logical relations with other tables. The organization structure table has a "one to many" relation which means that each position in the OBS can have several positions under it.

Although the construction industry still prefers traditional hard copy documentation [1], PMSS aims to encourage construction companies to use IT and Internet facilities by its motivation system. The system will be installed on the company's main server and each user can login through the Internet. The user interfaces are as user friendly as possible, so that there is no need to be a professional computer user. The developed system facilitates the company to keep a copy of every single piece of work in its databases.

# PMSS ENVIRONMENT

PMSS can be accessed through the Internet by different web browsers, such as Mozilla Firefox, Microsoft Internet Explorer, Google Chrome, Opera, etc. Users of the system can be anyone who is assigned a username and a password. The system consists of the following parts.

## OBS Development

The setup and installation authority is given to the company's General Manager (GM) to define the company OBS. Among the different organizational structures [82], a functional organizational structure is used in PMSS, since most of the companies prefer it nowadays [83]. OBS creation has a dynamic structure which means that there is no limit to the number of horizontal and vertical levels; thus, each company can create its own OBS structure. One of the weaknesses of the functional structure is its low team member motivation [83]. PMSS, with its performance measurement feature and its motivation system, is deemed well to overcome this problem and make the functional structure more efficient. GM can delegate his authority to other personnel, such as head of the estimating and tendering department or head of the design department, to be able to create, define or modify projects, activities and resources. Other users have only access to the data for their work. A sample page of OBS creation is shown in Figure 2.

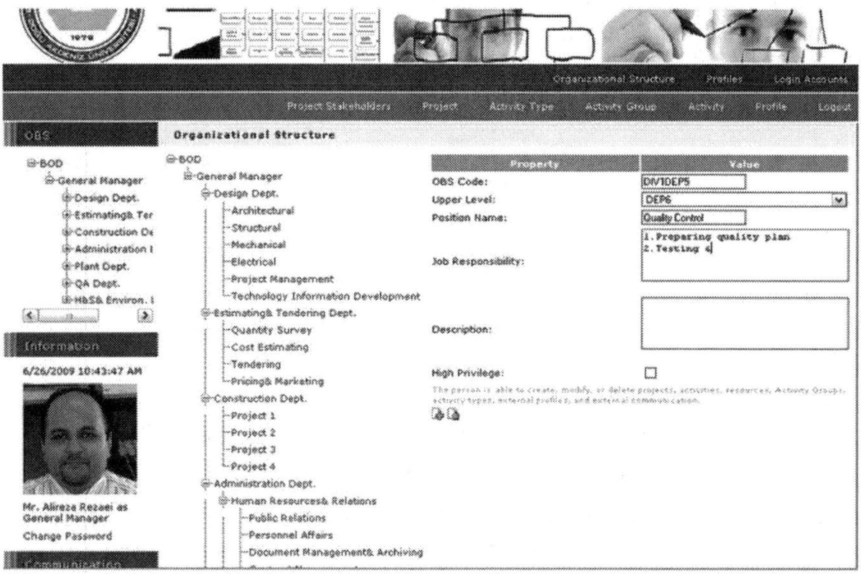


**Figure 2:** OBS creation in PMSS.

## Project, Activity and Resource Definition

Projects, activities and resources are specified by estimating and tendering department. After defining the projects and their related activities and resources, the system is ready to be implemented. Projects are defined to the system using the interface shown in Figure 3.

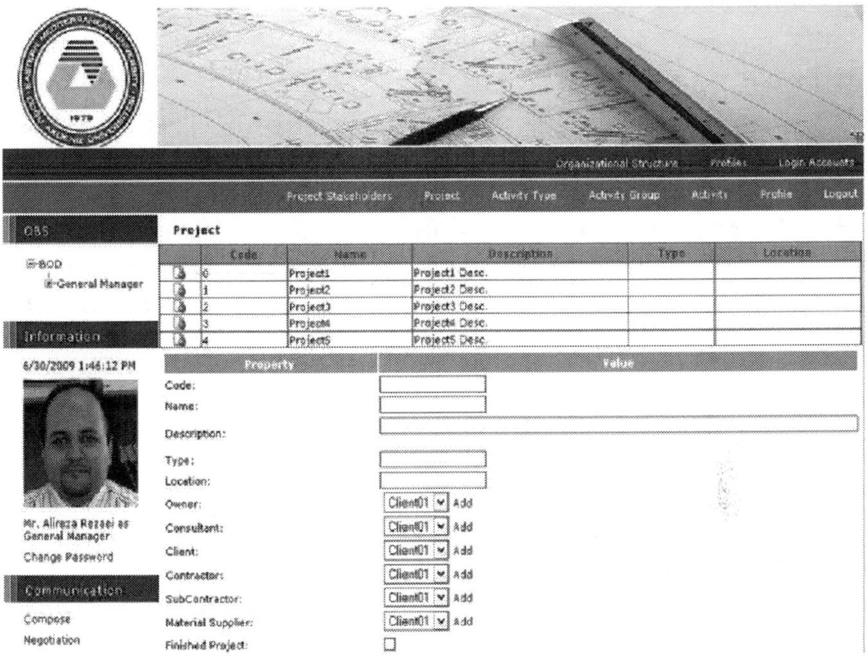

**Figure 3:** Project definition in PMSS.

## Message Transfer System

All sending and receiving of work in PMSS are done through message transactions between users of the system, and through predefined lines of communication. Each user has a communication panel including "Negotiation", "Submitted" and "Performed" inboxes in his interface, which allows composing a message, negotiating over a received work, submitting work or performing work. Each message has two parties: a sender and a recipient. The sender of the message has alternatives of "Negotiate" on resources of the work or "Submit" it directly to the recipient. If the message came to the recipient through "Negotiate", he has the right to send back his opinion to the sender. This process can continue until both parties come to an agreement on the estimated resources required for the work.

If the work came from the sender through "Submit", then the recipient has no right to object to the estimated resources, just do the work and send it back. The resources can be time, money, machinery, material and manpower. The sender is capable of checking the workload of the recipient to fix the duration of the work.

The recipient of any message will be informed by a short message through his mobile phone as soon as receiving a message in his inbox. In a typical compose page, as shown in Figure 4, the head of the design department prepared a message to the head of the estimating and tendering department for estimating the price of a tower crane.

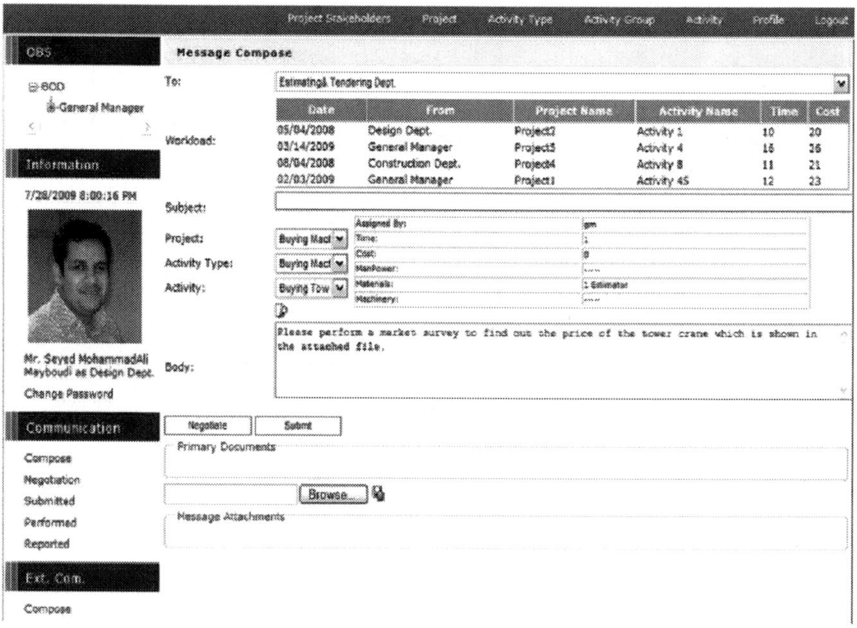


**Figure 4:** Compose page.

Once the work is completed by the recipient it will be sent back to the "performed" inbox of the original sender. Performed resource fields will be filled by the recipient, based on actual resources used. A sample of the performed message from the head of the estimating and tendering department to the head of the design department is shown in Figure 5.

**Figure 5:** Performing message in PMSS.

## Performance Factor Calculation

Once the original sender receives a performed work, he will compare the estimated resources with the actual resources used, check the accuracy of the work and assign a grade for its quality. For every performed work, an Activity Performance Factor (APF) will automatically be calculated by the system using Eq. (1).

Zegordi and Nahavandi's [84] simple definition of productivity is considered in Eq. (1). They defined productivity to be the ratio of output to input:

$$APF = \frac{T_{est}}{T_{per}} \times WT + \frac{C_{est}}{C_{per}} \times Wc + \frac{Q_{per}}{Q_{est}} \times W_Q \quad (1)$$

$_T est$ and $_C est$ are the estimated time and cost, respectively. $_Q est$ of all activities is always assumed to be 100 and is not negotiated. Hoehn [85] and Neely et al. [57] pointed out the need to recognize different weights when measuring performance. $_{WT'WC}$ and $_{WQ}$ represent the weighted factors of time, cost and quality, respectively. These factors are defined for every single activity, while specifying its resources by the estimating and tendering department. They explain the importance of each parameter. $_{WT'WC}$ and $_{WQ}$ are considered as percentages and their sum is always equal to 100%.

$_T per$ and $_C per$ represent the actual time and cost used. $_Q per$ is the quality grade and is assigned by the original sender on the basis of 0 to 100 measures. The subjectivity of $_Q per$ is believed to be minimized by considering the number of mistakes/errors/deficiencies, the delay in reading messages, and percentage of work actually completed in $_T per$ and $_C per$. One higher level authority will also be informed about the assigned $_Q per$. $_Q per$ will be calculated by the following formula:

$$Q_{per} = \frac{D_{est}}{D_{per}} \times W_D + \frac{M_{est}}{M_{per}} \times W_M + \frac{PC_{per}}{PC_{est}} \times W_{PC} \quad (2)$$

where $_D est$ is acceptable delay for the company and is assigned by the estimating and tendering department, $_D per$ is recipient delay time in reading the message, which is a percentage of $_D est$, $_M est$ is the number of allowed mistakes/errors/deficiencies for a specific activity, and varies for different activities, which is assigned by the estimating and tendering department, $_M per$ is the number of mistakes/errors/deficiencies occurred in a performed work, and $P_C per$ is the percentage of work completed. $P_C est$ is the estimated percentage of work completed within the assigned time ($T_{est}$) and by default is assumed to be 100%. $_{WD'WM}$ and $_{WPC}$ are weighted factors of delay, mistakes/errors/deficiencies and percentage of work completed, respectively. $_{WD'WM}$ and $_{WPC}$ are considered as percentages and defined by the estimating and tendering department for each activity, and their sum is always equal to 100%.

APF may be equal, less or more than 100%. If it is less than 100%, it means that the submitted work was performed at a lower level than expected. If it is equal to 100%, it means that it was performed as expected. If it is greater than 100%, it means that it was performed at a higher level than expected.

Every employee has a Performance Factor (PF) in the company. PF is updated, based on the OBS level, by using Eq. (3) or (4).

If the employee is at the lowest level in OBS, nobody gives him reports:

$$PF = \frac{\sum_{1}^{n} APFi}{n} \times (1 - 0.005w) \qquad (3)$$

Where APFi is the work performance obtained from a specific work and n is the number of total work performed so far, and w is the work experience of the employee in years. Other factors like the salary and education of an employee could also impact PF; however, only work experience was considered due to its higher importance.

If there are lower positions in OBS, giving reports to him:

$$PF = \frac{\frac{\sum_{1}^{n} APFi}{n} \times (1 - 0.005w) + \frac{\sum_{1}^{m} PFj}{m}}{2} \qquad (4)$$

Where PFj is the performance factor of every lower level employee, and m is the number of lower level employees that so far give reports to him.

Obviously, an employee with a higher PF will have better standing in the company. Besides, depending on the policy of the company, these factors can be effective in determining an increment to the salary or other types of reward, since rewards will support the performance measurement system in being more effective [86] and [87]. Performance measurement can actually change the lives of people and organizations if implemented and used properly [88]. Thus, it is expected that such an application will strongly motivate employees to perform better.

## PMSS Reports

DW can extract the data from different departments and projects and, after cleansing, use them to provide various reports. It can also keep them as historical data for future use and reference. For instance, PFs can be shown on the OBS chart, as one of the reports of the system, as shown in Figure 6. The reports are online and updated automatically as any change occurs in the work and database.

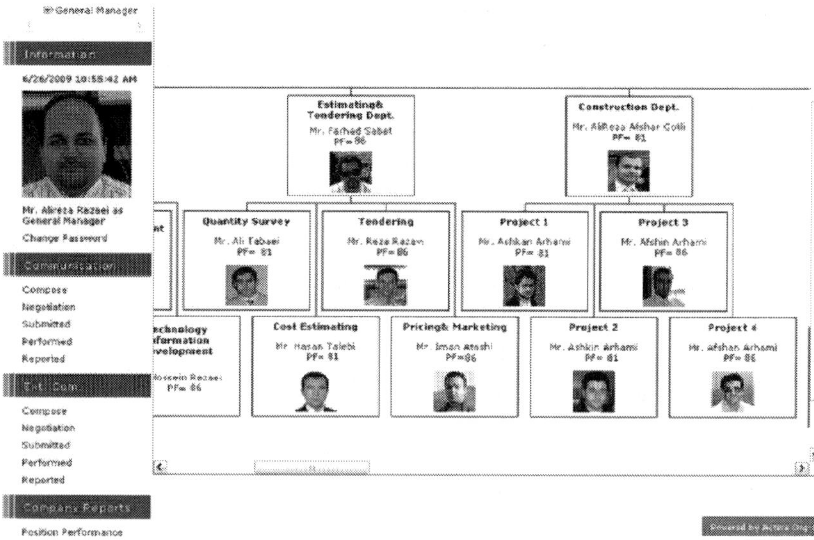


**Figure 6:** OBS with PFs report.

Another report PMSS provides is the amount of delay each person caused in message reading. The difference between receiving and reading times of all received messages will be calculated and accumulated for each person. These delays could be shown both in tabular format and graphically, and could also be shown on the OBS chart. Cost reports, cash flow reports, activity reports, position reports and many other types of reports could also be retrieved based on need.

# SYSTEM EVALUATION

As stated by Laitinen et al. [89], IS can provide several benefits, such as improved productivity, innovation, market share and cost saving. The evaluation of an IS can be viewed as an exercise in determining its implementation success. Implementation success refers to different factors to be achieved; thus, evaluation is a multidimensional concept [90]. These factors vary according to different group (users, developers, managers) perspectives and the type of information system to be evaluated.

## Primary System Evaluation

Since the main focus of PMSS is construction companies, the pilot model was run with managers of some construction companies in Northern Cyprus and Iran, and ISO 9000 certification agencies in Iran.

The primary feedbacks attested that PMSS will provide a competitive environment among employees and thus can help companies to overcome the difficulties of quality management system implementation and ISO certification. Managers pointed out some benefits provided by PMSS including time saving, cost and expense reduction (as a result of decreased paperwork, reduction of administrative staff, decrease in office expenses, decrease in archiving space, etc.) accuracy increase, easier and faster performance measurement and calculation, and communication improvement. However, there were some recommendations for improving the practicality of the system, including:

- Allowing a time extension after initial submission of the work;
- Considering matrix OBS as well;
- Considering three other resources (material, manpower, and machinery) in performance calculation as separate parameters;
- Checking the recipient workload before submitting new work.

The feedbacks of the managers were then analyzed and most of them applied to the system. A time extension, after fixing the estimated resources, was added to the system, so that it was possible to bargain and change the resources after work submission. The matrix OBS could not be considered in the system since it is considered a limitation.

Material, manpower and machinery were already considered to affect the cost, and their affects were included in the $_{C}\text{est}$ and $_{C}\text{per}$. The recipient's existing workload consideration was also included, as shown in Figure 4. The modified PMSS was run again with the same managers, and they found it satisfactory and recommendable.

## Final System Evaluation

After final revision, the modified PMSS was practically installed in two construction companies in Iran at the end of 2009. After almost one month of training and debugging the system, it was implemented in those companies. The authors planned for two different feedbacks including short-term and long-term; the former from managers and the latter from both managers and employees. Short-term feedback was attained after one month of practical implementation. The summary of the short-term feedback from managers of two construction companies is as follows:

### *Cost Reduction*

One of the main advantages of PMSS was stated to be cost reduction. The managers believed that reducing the number of administrative personnel up to 50% would lead the company to cost savings including salaries, office space, furniture, indirect costs, food, etc. Another benefit of reducing the number of employees was stated to be making management more straightforward. It is obvious that managing a company with less numbers of employees is much easier. Cost saving and streamlining management would be more tangible for large companies.

### *Time Saving*

Another advantage was reported as being time saving. As declared by managers, the average delivery time for different work would be minimized by a message transfer system. It was also highlighted by ISO certification agencies that according to their experience with other IT systems, the spent administrative time would reduce between 40% and 60%.

## Other Benefits

Less paperwork, proper documentation and archiving and easier communication were also emphasized by managers, as a result of DW and IT usage. One manager expressed that archiving space, after ISO 9000 certification, was expanded to being 4 times bigger than before certification. They had to extend the buildings for administrative and archiving purposes. This imposed considerable expense on the company, including the need for new personnel, and also caused some difficulties in communication and data access. He claimed that using a system like PMSS would help companies prevent such extra expense and effort. On the other hand, an important issue for a manager is fast and easy access to necessary information, which can be achieved easily by utilizing a system like PMSS.

Managers also revealed that more accuracy would be attained by using PMSS. Almost 100% accuracy would be provided, compared with 15%–20% human errors faced by companies as stated by some managers.

## Application Difficulties

The only difficulty of PMSS application was believed to be the resistance of personnel. Clearly, every new system faces some resistance and adaptation difficulties. Management commitment and ability would encourage personnel to accept the new system and procedures.

Long-term feedback was expected to be obtained after longer time usage of PMSS. The authors expect to receive both managers and employees feedback after at least one year of PMSS practical implementation. As short-term feedback demonstrated the practicality and advantages of PMSS, together with the positive view of the managers, it is anticipated that long-term feedback will also prove its appropriateness and ease of use. Motivation cannot be measured yet, since it requires employee performance to be observed and evaluated after at least 6 months of PMSS implementation. However, with the effects of PFs on salaries and recognition, motivation will undoubtedly be imposed on employees. Therefore, PMSS advantages will be more documented with long-term feedback results.

# PRACTICAL AND MANAGERIAL IMPLICATIONS OF PMSS

Performance measurement systems and information systems receive considerable attention nowadays; thus it is expected that PMSS, with both characteristics, will provide numerous and important advantages for its users. Although PMSS is still running in two construction companies, and long-term feedback has not yet been received, it is believed, and somehow proved by short-term feedback results, that some main practical and managerial implications of PMSS are:

- Positive effects on working behavior of employees and managers.
- Creation of a competitive environment among employees to perform their work in the shortest time, with the lowest cost, and to the highest quality level.
- Provision of a decision tool for managers to have quick and reliable access to employee performance, and to monitor the overall performance of their organization.
- Motivation of employees for better performance as a human resource management tool.
- To encourage companies seeking ISO 9000 certification or QMS deployment to start and continue the process, and overcome implementation difficulties.
- Reduction of costs through a reduction in the number of administrative personnel, office space and equipment, archiving space, etc.
- Assist organizations to make the best use of resources, such as time, cost, manpower, machinery and materials.

# CONCLUSIONS

Companies seeking quality management system deployment, like ISO 9000 certification, face numerous discouraging difficulties. Major difficulties are paperwork, documentation, communication and lack of motivation.

Using a web-based office automation system, called Performance Management Support System (PMSS), managers would be promptly provided with accurate information about their business. PMSS reduces paperwork through appropriate IT utilization, decreases documentation problems by the use of a data warehouse, overcomes communication problems by using Internet and mobile phone facilities, and, finally, assists the company in getting ISO certification as a part of its quality management system.

On the other hand, PMSS encourages employees to perform their jobs in the possible shortest time, with the lowest cost, and at the highest quality level to gain higher PFs. Obtaining higher PFs will provide a better standing of the employee in the company and contribute to earning more prerequisites.

PMSS only considers functional OBS and takes three indexes of time, cost and quality in performance calculations. Further studies can be conducted on other types of OBS and also separate consideration of three other resources in calculations.

# REFERENCES


1. M. Alshawi, B. Ingirige Web-enabled project management: an emerging paradigm in construction Automation in Construction, 12 (2003), pp. 349–364
2. K.C. Lam, S.T. Ng A cooperative Internet-facilitated quality management environment for construction Automation in Construction, 15 (2006), pp. 1–11
3. R. Sroufe, S. Curkovic An examination of ISO 9000: 2000 and supply chain quality assurance Journal of Operations Management, 26 (2008), pp. 503–520
4. T. Ng Performance of engineering consultants in ISO 9000-based quality management systems implementation Engineering, Construction and Architectural Management, 12 (6) (2005), pp. 519–532
5. L.S. Pheng, W. Wee Improving maintenance and reducing building defects through ISO 9000 Journal of Quality in Maintenance Engineering, 7 (1) (2001), pp. 6–24

6. S. Allesina, A. Azzi, D. Battini, A. Regattieri Performance measurement in supply chains: new network analysis and entropic indexes International Journal of Production Research, 48 (8) (2010), pp. 2297–2321
7. M. Pero, T. Rossi, C. Noe, A. Sianesi An exploratory study of the relation between supply chain topological features and supply chain performance International Journal of Production Economics, 123 (2010), pp. 266–278
8. L. Whicker, M. Bernon, S. Templar, C. Mena Understanding the relationships between time and cost to improve supply chain performance International Journal of Production Economics, 121 (2009), pp. 641–650
9. S.W. Kim An investigation on the direct and indirect effect of supply chain integration on firm performance International Journal of Production Economics, 119 (2009), pp. 328–346
10. C.W. Craighead, G.T.M. Hult, D.J. Ketchen Jr. The effects of innovation-cost strategy, knowledge, and action in the supply chain on firm performance Journal of Operations Management, 27 (2009), pp. 405–421
11. A.C.L. Yeung Strategic supply management, quality initiatives, and organizational performance Journal of Operations Management, 26 (2008), pp. 490–502
12. M. Kojima, K. Nakashima, K. Ohno Performance evaluation of SCM in JIT environment International Journal of Production Economics, 115 (2008), pp. 439–443
13. P.D. Cousins, B. Lawson, B. Squire Performance measurement in strategic buyer–supplier relationships—the mediating role of socialization mechanisms International Journal of Operations and Production Management, 28 (3) (2008), pp. 238–258
14. A.M. Sharif, Z. Irani, D. Lloyd Information technology and performance management for build-to-order supply chains International Journal of Operations and Production Management, 27 (11) (2007), pp. 1235–1253
15. S. Carr, H. Kaynak Communication methods, information sharing, supplier development and performance—an empirical study of their relationships International Journal of Operations and Production Management, 27 (4) (2007), pp. 346–370

16. W.P. Wong, K.Y. Wong Supply chain performance measurement system using DEA modeling Industrial Management + Data Systems, 107 (3) (2007), pp. 361–381
17. M. Sepehri, K. Fayazbakhsh, F. Ghasemzadeh A corporate supply optimizer with flow network Scientia Iranica, Transactions E: Industrial Engineering, 17 (1) (2010), pp. 70–83
18. .R. Hosseini Baharanchi Investigation of the impact of supply chain integration on product innovation and quality Scientia Iranica, Transactions E: Industrial Engineering, 16 (1) (2009), pp. 81–89
19. M.H. Fazel Zarandi, I.B. Turksen, S.M. Moattar Hoseini, S. Bastani, A. Mohebi A fuzzy intelligent information agent architecture for supply chains Scientia Iranica, 15 (5) (2008), pp. 623–636
20. K. Phusavat, P. Anussornnitisarn, S. Sujitwanit, P. Kess Profile-based circumstances for productivity measurement Industrial Management + Data Systems, 109 (6) (2009), pp. 825–839
21. L.A. Ho What affects organizational performance? The linking of learning and knowledge management Industrial Management + Data Systems, 108 (9) (2008), pp. 1234–1254
22. C.Y. Lin, T.H. Kuo The mediate effect of learning and knowledge on organizational performance Industrial Management + Data Systems, 107 (7) (2007), pp. 1066–1083
23. M.P. Rao A performance measurement system using a profit-linked multi-factor measurement model Industrial Management + Data Systems, 106 (3) (2006), pp. 362–379
24. G.H. Saad Strategic performance evaluation: descriptive and prescriptive analysis Industrial Management + Data Systems, 101 (8) (2001), pp. 390–399
25. C.H. Kuo, K.D. Dunn, S.U. Randhawa A case study assessment of performance measurement in distribution centers Industrial Management + Data Systems, 99 (2) (1999), pp. 54–63
26. K. Arjomandi, H. Estekanchi, A. Vafai Correlation between structural performance levels and damage indexes in steel frames subjected to earthquakes Scientia Iranica, Transactions A: Civil Engineering, 16 (2) (2009), pp. 147–155
27. M. Mirabi, S.M.T. Fatemi Ghomi, F. Jolai A hybrid electromagnetism-like algorithm for supplier selection in make-to-order planning

Scientia Iranica, Transactions E: Industrial Engineering, 17 (1) (2010), pp. 1–11

28. S.M. Seyedhoseini, M.A. Hatefi Two-pillar risk management (TRPM): a generic project risk management process Scientia Iranica, Transactions E: Industrial Engineering, 16 (2) (2009), pp. 138–148
29. F. Jariri, S.H. Zegordi Quality function deployment, value engineering and target costing, an integrated framework in design cost management: a mathematical programming approach Scientia Iranica, 15 (3) (2008), pp. 405–411
30. A. Azaron, M. Modarres Project completion time in dynamic PERT networks with generating projects Scientia Iranica, 14 (1) (2007), pp. 56–63
31. S.T. Foster Jr. Towards an understanding of supply chain quality management Journal of Operations Management, 26 (2008), pp. 461–467
32. S. Chin, K. Kim, Y. Kim A process-based quality management information system Automation in Construction, 13 (2004), pp. 241–259
33. A.M. Turk ISO 9000 in construction: an examination of its application in Turkey Building and Environment, 41 (4) (2006), pp. 501–511
34. R. Motazed-Keyvani, A.R. Ghanbari-Parsa, S. Kagaya ISO 9000 standards: perceptions and experiences in the UK construction industry Construction Management and Economics, 17 (1999), pp. 107–119
35. Y.S. Kim, S.W. Oh, Y.K. Cho, J.W. Seo A PDA and wireless web-integrated system for quality inspection and defect management of apartment housing projects Automation in Construction, 17 (2008), pp. 163–179
36. V.W.Y. Tam, K.N. Le Quality improvement in construction by using a Vandermonde interpolation technique International Journal of Operations and Production Management, 25 (2007), pp. 815–823
37. M. Martinez-Costa, A.R. Martinez-Lorente, T.Y. Choi Simultaneous consideration of TQM and ISO 9000 on performance and

motivation: an empirical study of Spanish companies International Journal of Production Economics, 113 (2008), pp. 23–39

38. M. Karamouz, A. Ahmadi, S. Nazif Development of management schemes in irrigation planning: economic and crop pattern consideration Scientia Iranica, Transactions A: Civil Engineering, 16 (6) (2009), pp. 457–466
39. R. Mostaghel, A. Albadvi A cross-cultural comparison of service quality prioritization Scientia Iranica, Transactions E: Industrial Engineering, 16 (1) (2009), pp. 65–72
40. L.Y. Fok, S.J. Hartman, A.L. Patti, J.R. Razek Human factors affecting the acceptance of total quality management International Journal of Quality & Reliability Management, 17 (7) (2000), pp. 714–729
41. A. Neely, M. Gregory, K. Platts Performance measurement system design—a literature review and research agenda International Journal of Operations and Production Management, 15 (4) (1995), pp. 80–116
42. A. Neely, M. Gregory, K. Platts Performance measurement system design—a literature review and research agenda International Journal of Operations and Production Management, 25 (12) (2005), pp. 1228–1263
43. K. Phusavat, P. Anussornnitisarn, P. Helo, R. Dwight Performance measurement: roles and challenges Industrial Management + Data Systems, 109 (5) (2009), pp. 646–664
44. K. Phusavat, W. Photaranon Productivity/performance measurement, case application at the government pharmaceutical organization Industrial Management + Data Systems, 106 (9) (2006), pp. 1271–1287
45. H.I. Kulmala, L. Ahoniemi, V. Nissinen Performance through measuring leader's profiles: an empirical study International Journal of Production Economics, 122 (2009), pp. 385–394
46. R. Johnston, P. Pongatichat Managing the tension between performance measurement and strategy: coping strategies International Journal of Operations and Production Management, 28 (10) (2008), pp. 941–967
47. M. Marchand, L. Raymond Researching performance measurement systems—an information systems perspective

International Journal of Operations and Production Management, 28 (7) (2008), pp. 663–686

48. P. Garengo, U. Bititci Towards a contingency approach to performance measurement: an empirical study in Scottish SMEs International Journal of Operations and Production Management, 27 (8) (2007), pp. 802–825
49. J. Ukko, J. Tenhunen, H. Rantanen Performance measurement impacts on management and leadership: perspectives of management and employees International Journal of Production Economics, 110 (2007), pp. 39–51
50. U.S. Bititci, K. Mendibil, S. Nudurupati, P. Garengo, T. Turner Dynamics of performance measurement and organizational culture International Journal of Operations and Production Management, 26 (12) (2006), pp. 1325–1350
51. A. Neely The evolution of performance measurement research—developments in the last decade and a research agenda for the next International Journal of Operations and Production Management, 25 (12) (2005), pp. 1264–1277
52. U.S. Bititci, K. Mendibil, V. Martinez, P. Albores Measuring and managing performance in extended enterprises International Journal of Operations and Production Management, 25 (4) (2005), pp. 333–353
53. J.R. Evans An exploratory study of performance measurement systems and relationships with performance results Journal of Operations Management, 22 (2004), pp. 219–232
54. U.S. Bititci, S.S. Nudurupati, T.J. Turner, S. Creighton Web enabled performance measurement systems—management implications International Journal of Operations and Production Management, 22 (11) (2002), pp. 1273–1287
55. M. Kennerley, A. Neely A framework of the factors affecting the evolution of performance measurement systems International Journal of Operations and Production Management, 22 (11) (2002), pp. 1222–1245
56. M. Bourne, J. Mills, M. Wilcox, A. Neely, K. Platts Designing, implementing and updating performance measurement systems International Journal of Operations and Production Management, 20 (7) (2000), pp. 754–771

57. A. Neely, J. Mills, K. Platts, H. Richards, M. Gregory, M. Bourne, M. Kennerley Performance measurement system design: developing and testing a process-based approach International Journal of Operations and Production Management, 20 (10) (2000), pp. 1119–1145
58. K.W. Green Jr., B. Medlin, D. Whitten Developing optimism to improve performance: an approach for the manufacturing sector Industrial Management + Data Systems, 104 (2) (2004), pp. 106–114
59. B. Medlin, K.W. Green Jr. Enhancing performance through goal setting, engagement, and optimism Industrial Management + Data Systems, 109 (7) (2009), pp. 943–956
60. M.Y. Tarng, H.Y. Liu Creating a document management system Industrial Management + Data Systems, 94 (9) (1994), pp. 9–15
61. E.K. Laitinen Importance of performance information in managerial work Industrial Management + Data Systems, 109 (4) (2009), pp. 550–569
62. U. Ojiako, E. Johansen, D. Greenwood A qualitative re-construction of project measurement criteria Industrial Management + Data Systems, 108 (3) (2008), pp. 405–417
63. S. Ahmad, R.G. Schroeder The impact of human resource management practices on operational performance: recognizing country and industry differences Journal of Operations Management, 21 (2003), pp. 19–43
64. A.Y. Nahm, M.A. Vonderembse, X.A. Koufteros The impact of organizational structure on time-based manufacturing and plant performance Journal of Operations Management, 21 (2003), pp. 281–306
65. A. De Toni, S. Tonchia Performance measurement systems—models, characteristics and measures International Journal of Operations and Production Management, 21 (1–2) (2001),
66. M. Najmi, D.F. Kehoe The role of performance measurement systems in promoting quality development beyond ISO 9000 International Journal of Operations and Production Management, 21 (1–2) (2001), pp. 159–172

67. J. Jayaram, C. Droge, S.K. Vickery The impact of human resource management practices on manufacturing performance Journal of Operations Management, 18 (1999), pp. 1–20
68. R. Bose Understanding management data systems for enterprise performance management Industrial Management + Data Systems, 106 (1) (2006), pp. 43–59
69. S.T. March, A.R. Hevner Integrated decision support systems: a data warehousing perspective Decision Support Systems, 43 (2007), pp. 1031–1043
70. G.M. Marakas Systems Analysis & Design—an Active Approach (2nd ed.)McGraw-Hill, USA (2006)
71. H. Golnabi, P. Azimi, G.H. Poseh Data acquisition and resistance measurements using a simple program Scientia Iranica, 14 (5) (2007), pp. 467–473
72. L. Hedelin, C.M. Allwood IT and strategic decision making Industrial Management + Data Systems, 102 (2) (2002), pp. 125–139
73. M. Jarke, M. Lenzerini, Y. Vassiliou, P. Vassiliadis Fundamentals of Data Warehouses Springer, Cambridge, MA (2000)
74. .C.W. Lau, F.T.S. Chan, R. Fung, C.W.Y. Wong An XML-based real-time quality measurement scheme Industrial Management + Data Systems, 104 (6) (2004), pp. 505–512
75. S.M. Lee, S. Hong An enterprise-wide knowledge management system infrastructure Industrial Management + Data Systems, 102 (1) (2002), pp. 17–25
76. I. Ahmad, S. Azhar, P. Lukauskis Development of a decision support system using data warehousing to assist builders/developers in site selection Automation in Construction, 13 (2004), pp. 525–542
77. M. Zhiliang, K.D. Wong, L. Heng, Y. Jun Utilizing exchanged documents in construction projects for decision support based on data warehousing techniques Automation in Construction, 14 (2005), pp. 405–412
78. S.W. Moon, J.S. Kim, K.N. Kwon Effectiveness of OLAP-based cost data management in construction cost estimate Automation in Construction, 16 (2007), pp. 336–344

79. N. Forcada, M. Casals, X. Roca, M. Gangolells Adoption of web databases for document management in SMEs of the construction sector in Spain Automation in Construction, 16 (2007), pp. 411–424
80. J. Ang, T.S.H. Teo Management issues in data warehousing: insights form the housing and development board Decision Support Systems, 29 (2000), pp. 11–20
81. R. Lair Beginning Silverlight 2 A Press, New York, USA (2008)
82. C.J. Anumba, C. Baugh, M.M.A. Khalfan Organizational structures to support concurrent engineering in construction Industrial Management + Data Systems, 102 (5) (2002), pp. 260–270
83. J.K. Pinto Project Management. Achieving Competitive Advantage Pearson Educations International, New Jersey, USA (2007)
84. S.H. Zegordi, N. Nahavandi Measuring productivity indexes and efficiency in the container terminal at port Rajaei Scientia Iranica, 9 (3) (2002), pp. 248–254
85. Hoehn, W. "Managing organizational performance: linking the balanced scorecard to a process improvement technique", *Proc. 4th Ann. Int. Symp. Ind. Eng. Per. Manag.* Kasetsart University, Bangkok, pp. 1–12 (2003).
86. V. Hanna, N.D. Burns, C.J. Backhouse Realigning organizational variables to support workplace behaviour International Journal of Operations and Production Management, 20 (12) (2000), pp. 1380–1391
87. M.E. Kuwaiti Performance measurement process: definition and ownership International Journal of Operations and Production Management, 24 (1) (2004), pp. 55–78
88. K.Y. Kutucuoglu, J. Hamali, Z. Irani, J.M. Sharp A framework for managing maintenance using performance measurement systems International Journal of Operations and Production Management, 21 (1–2) (2001), pp. 173–194
89. E.K. Laitinen, A. Lansiluoto, I. Rautiainen Extracting appropriate scope for information systems: a case study Industrial Management + Data Systems, 109 (3) (2009), pp. 305–321

90. S.C. Palvia, R.S. Sharma, D.W. Conrath A socio-technical framework for quality assessment of computer information systems Industrial Management + Data Systems, 101 (5) (2001), pp. 237–251

Chapter 5

# Engagement of Facilities Management in Design Stage Through Bim: Framework and a Case Study

Ying Wang[1], Xiangyu Wang[1], Jun Wang[2], Ping Yung[1], and Guo Jun[3]

[1]School of Built Environment, Curtin University of Western Australia, Australia

[2]School of Construction Management and Real Estate, Chongqing University, China

[3]CCDI, China

## ABSTRACT

Considering facilities management (FM) at the early design stage could potentially reduce the efforts for maintenance during the operational phase of facilities. Few efforts in construction industry have involved facility managers into the design phase. It was suggested that early adoption of facilities management will contribute to reducing the needs for major repairs and alternations that will otherwise occur at the operational phase. There should be an integrated data source providing information support for the building lifecycle. It is envisaged that Building Information Modelling (BIM) would fill the gap by acting as a visual model and a database throughout the building lifecycle. This paper develops a framework of how FM can be considered in design stage through BIM. Based on the framework, the paper explores how BIM will beneficially support FM in the design phase, such as space planning and energy analysis. A case study of using BIM to design facility managers' travelling path in the maintenance process is presented. The results show that early adoption of FM in design stage with BIM can significantly reduce life cycle costs.

# INTRODUCTION

According to the International Facility Management Association (IFMA), facility management (FM) is defined as "a profession that encompasses multiple disciplines to ensure functionality of the built environment by integrating people, place, processes and technology" [1]. Industries in varieties of areas are adopting BIM for FM. Organizations such university, government, healthcare, retail, and information technology are taking a survey for the adoption of BIM-based FM [2]. Different parts of FM are adopted with BIM in these organizations. Figure 1 depicts the proportion of each function.

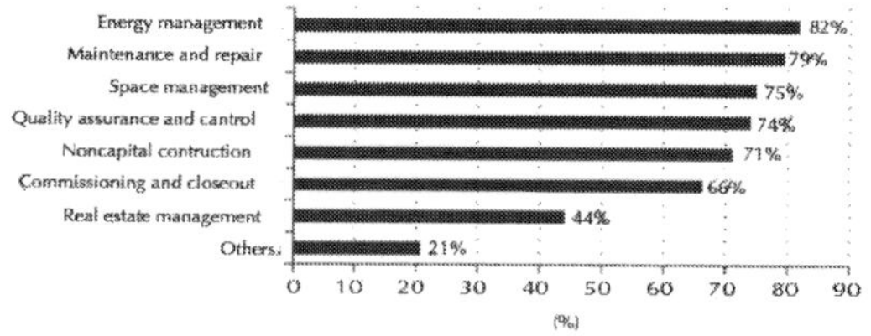


**Figure 1:** Proportion of each function [2].

However, few efforts in the construction industry have involved facility FM into the design phase [3, 4]. It was suggested that early engagement of FM would contribute to reducing the needs for major repairs and alternations that will otherwise occur at the operational phase [2, 5]. There have been rare effective approaches or processes to engage FM in design stage. The proposed framework of this paper is going to integrate these FM works into early design stage which could potentially strengthen the collaboration between design team and FM team and reduce alternations. BIM is envisaged to be an effective tool, as proposed in this paper. Considering the multidisciplinary and interoperability of this process, there must be a data source providing convenient integration and access to the relevant information. Building Information Modelling (BIM) is a conceptual approach to building design and construction that comprises all the graphic and linguistic data of building for design and detailing which facilitates exchange of building information between design, construction, and operational phase [6]. A BIM model could comprise individual 3D models of each building component with all associated properties such as weight, material, length, height, geographical information system GIS and information [7]. Beyond the inherent information, BIM also includes external association between building components. For example, the column with name Col. –093 is installed between box ceiling Cei. –52 and level 2 floor with GUID number 30836. Figure 2 depicts the column model and associated properties. The main difference between BIM and 2D CAD is that the latter describes a building layer by layer [8]. Editing one layer will result in massive updating and checking work of associated floor plan. In contrast, BIM models are designed in terms of building components in 3D view. An error-prone process such as clash detection could be conducted automatically [9]. This paper aims to develop a framework of bringing facility management into design stage through BIM. Which field of FM work should be brought to the design stage and which data should be collected are also proposed. With the ease of access to lifecycle information of all the building components BIM provided, proposed building plan could be optimized and lifecycle cost could be reduced with the FM knowledge and experience.

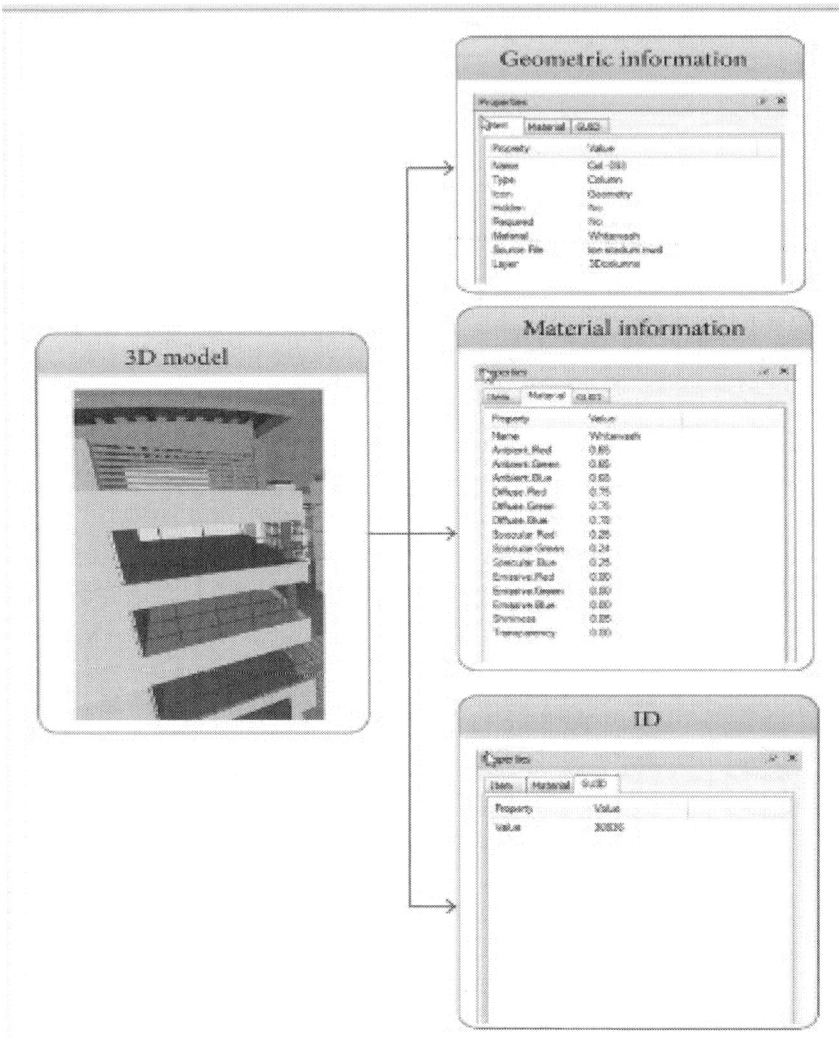


**Figure 2:** Column model and properties in BIM.

## Methodology: Conceptual Framework of Integrating FM and BIM in Design Stage

Erdener [5] developed a framework linking design with FM by programming—an extension of "problem defining-solving method" which however did not classify the specific field of work in FM that should be involved into the design stage. Additionally, the backend database was not adopted as an approach to integrate the massive information such as asset portfolios, instructions, and design manuals in this multidisciplinary process. Mostly, the operation and maintenance process of a facility occupies more than 80% in its lifecycle for both cost and time [10]. During the FM process, facility managers have to acquire, integrate, edit, and update massive information related to diverse building elements such as operational costs, warranties, and specifications from varieties of systems. BIM could effectively merge these primary data and provide convenient storage and retrieval of these FM data. Based on the work of Becerik-Gerber et al. [2], three types of FM data should be incorporated into BIM: (1) equipment and systems, (2) attributes and data, (3) portfolios and documents. Figure 3 illustrates the structure of the proposed BIM database for FM. Every facility in buildings is regarded as an individual entity with two kinds of properties—attributes and portfolios. Six types of basic equipment such as HVAC, plumbing, and electrical are represented as entities in BIM. Each entity has its attributes (vender information, location information, etc.) and attached documents (specifications, warranties, manuals, etc.). Specifically, serial numbers of products specified by vendors will be collected as unique identifier for each facility. Model and part numbers will act as reference information during the maintenance. Location information is comprised of building number, floor, and room number. Description stores the status of the facility. Attributes include weight, power, and energy consumption. In order to integrate the whole information into one standardised BIM database, interoperability needs to be assured. This is partially because in different circumstances different FM software systems are adopted, that is, Mainpac for building maintenance; FaPI for monitoring building condition; TRIM for document management. Nevertheless, all these software have their own data structure and usually they are not compatible with each other. International Alliance for Interoperability (IAI) published the Industry Foundation Classes (IFC), a standard for BIM data structure based on an ISO standard (ISO, 1994) enabling exchange of information among heterogeneous systems [11].

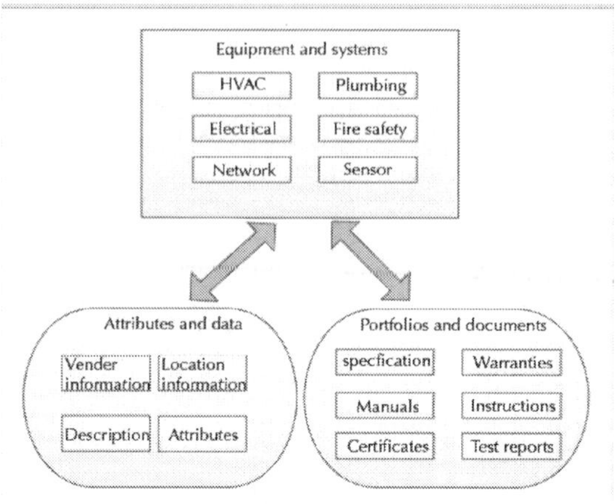


**Figure 3:** FM based BIM database.

BIM will provide supporting information for many categories of FM work such as maintenance and repair, energy management, commissioning, safety, and space management. Three categories of FM during the building's lifecycle are determined to be the most proper and specially discussed in this paper—(1) maintenance and repair, (2) energy management, and (3) commissioning. The decisions made in the design stage affect all aspects in maintenance stage and vice versa. The designer's relationship with the other participants in maintenance stages is very important [12]. Therefore, the maintenance team should also be involved into design stage for decision making. Additionally, different energy saving alternatives can be explored and simulated in early design stage with BIM [13]. Last but not least, commissioning stage ensures that a new building or system begins its life cycle at optimal productivity [14], in which coordination and information sharing between designer and participants essential.

This transformation will provide evaluation information for the design team and make the decision making much easier in both strategic-tactical and operational phase.

For the former, the facility manager could provide post occupancy evaluation of facilities for the design team as feedback. For the latter, bringing these FM jobs into the design stage will avoid redesign and reduce the maintenance job. The following subsection discusses the BIM role in FM engagement in design stage in detail.

## Maintenance and Repair

Maintenance is defined as activities required keeping a facility in as-built condition, while continuing to maintain its original productivity [12]. During this procedure, FM personnel have to identify the components' location and get access to the relevant documents, and finally, the maintenance information. In the state-of-the-art design phase, facility management relevant information such as working space of equipment, storage condition, and weight are not considered. This directly leads to the inappropriate allocation of space and incorrect estimation of load expectations.

Location information of facility could help facility managers efficiently identify the location of specific building components, especially for those who outsource the FM tasks. The knowledge and experience of facility managers could inform the architecture designers with working condition and space of different facilities. Both interior and exterior space requirements must be considered for the normal installation and implementation. Interior space refers to the working space, storage space, and privacy of the space. Exterior space includes the spaces needed for installation and, in case of emergency, for people's escape route. All the above issues could be incorporated into BIM and shown in graphical interface for the discussion between designers and facility managers.

Additionally, FM personnel could retrieve the relevant data of task from BIM's graphical interface in real time. For example, when troubleshooting a printer, FM personnel have to check the maintenance history, get the maintenance manual, generate maintenance reports, and close the request. Conventionally, they have to log on to different electronic document management systems (EDMSs) and toggle between multiple databases to retrieve relevant information.

Preventative maintenance (PM) is defined as the care and servicing by personnel for the purpose of maintaining equipment and facilities in satisfactory operating condition by providing for systematic inspection, detection, and correction of incipient failures either before they occur or before they develop into major defects [15]. For the matter of regular inspections, a schedule will be prepared. Detailed work description is preferred for improving the overall productivity that is work order ID, facility ID, location, description of the preventive work, documents required to perform maintenance, estimated and actual labour hours, and frequency of maintenance work [12]. All these data could be incorporated into BIM database as attributes and documents. Considering the unique ID of every facility, each one could be assigned an associated barcode for the ease of access to relevant information in real time through mobile device. Additionally, after every time of maintenance, status information and working hours will be sent to BIM as feedback and reference for next turn. Figure 4 depicts the workflow of BIM-based PM. Through predesigning of maintenance information such as location information, relevant maintenance history, and schedule for PM with BIM, incorporated information could be accessed conveniently. Future maintenance will be reasonable, and redesign is avoided.

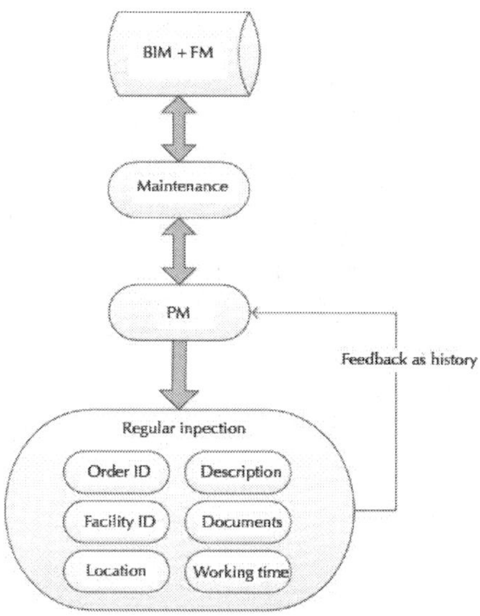


**Figure 4:** BIM-based PM workflow.

## Energy Management

Statistics from the US Green Building Council [16] show that in the United States, 72% of electricity consumption, 39% of energy use, and 38% of all carbon dioxide are from buildings. However, most buildings are not optimized in terms of energy consumption or not professionally optimized with advice or knowledge of facility management teams [13]. Torcellini et al. [17] identified "designing and constructing low-energy buildings (buildings that consume 50% to 70% less energy than code-compliant buildings) require the design team to follow an energy-design process that considers how the building envelope and systems work together." Energy consumption design must be set by the design team in the predesign phase. Afterwards, virtual prototyping will be created to simulate the energy efficiency. Acquainted with knowledge on energy codes and standards, the building energy consultant in FM team can provide all the information related to energy consumption for the basic energy analyses [17]. However, traditionally, most building energy analyses have been conducted late in design. Due to the difficulty and expense of modelling the energy systems, identifying and validating energy saving alternatives with different models is not economically possible. A large portion of time will be consumed in converting floor plans to energy management system graphics [4]. BIM is envisaged to be the platform for data exchanging avoiding reentering all the building geometry, enclosure, and HVAC information. Interoperability can be overcome by the data exchange standard gbXML (Green Building Extensible Markup). BIM software such as Bentley, Autodesk Revit, Graphisoft ArchiCAD, and Google sketchUpare able to export energy analyses data in gbXML format. For overall BIM energy design of a building, three steps have to be executed based on Kim et al.'s [13] work as follows.

1. Create BIM Model of Building. BIM model could be created based on the existing floor plan. This model comprises structured building components which include spatial data, texts, and databases of other properties. Based on these data, volume can be calculated, as well as energy estimates.
2. Integrate Energy Consumption Data and Test to Identify an Alternative. After modelling, energy consumption relevant data could be exported in gbXML format and analysed by tools such as Ecotect, Green Building Studio, and DOE-2. gbXML data could be used to analyse the energy consumption of the whole building, estimate water usage and cost evaluation, visualize solar radiation on surfaces, and simulate daylight factors [18]. Alternative design could be simulated by changing the lighting, roof, and walls.

3. Validation of the Proposed Design. After a design alternative is specified, validation from energy consultant is essential. Logic and assumptions of the energy model must be carefully reviewed. Figure 5 depicts the framework of energy management design using BIM.

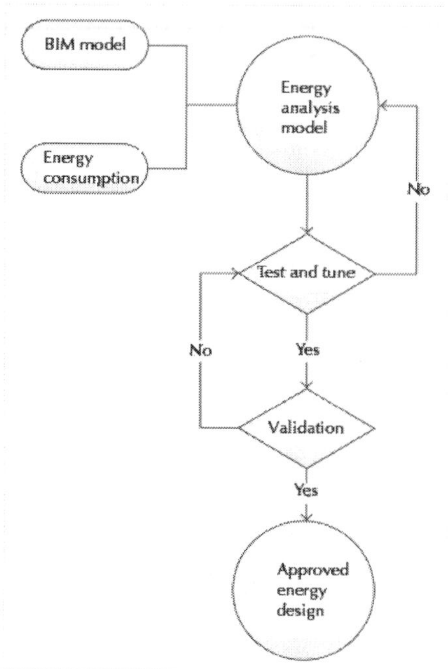


**Figure 5:** Framework of energy management design using BIM.

Treating each energy-consuming object as an entity, real time and period energy consumption will be collected as one of its property. Thus, cost information of a room/zone could be calculated by adding up all the energy cost of energy-consuming objects in it. Some high-energy-consumption facilities' information could be predicted by the historical data.

## Commissioning and Handover

Building commissioning is defined as "a quality assurance program intended to demonstrate that the building is constructed well and performs as designed. If the building materials, equipment and systems were not installed well and are not operating as intended, the health, productivity and other benefits of high performance design will not be achieved" [19]. Building commissioning is a key process for the building operation and maintenance, as the Department of Energy (DOE) suggested that "it ensures that a new building or system begins its life cycle at optimal productivity and improves the likelihood that the equipment will maintain this level of performance throughout its life. Building commissioning is the key to quality assurance in more than one way" [20]. For evaluating the project quality and identify potential significant design defects before it is too late or expensive to make changes, building commissioning should be embedded in the following phases: predesign phase, design phase, construction phase, transition to operational sustainability, postoccupancy and warranty phase, and retrocommissioning [21]. This section is focused on the first two processes: predesign phase and design phase. In design stages, commissioning scope and commissioning team must be identified. Since different facilities have different features and budget limitations and different projects have special systems to be commissioned, commissioning team has to be involved in early design stages for the decision making. This approach will also enable knowledge sharing between different parties. However, in the process of commissioning, massive 2D documents, images, maintenance, and operation information need to be collected and accessed frequently. For example, in the commissioning process of the Maryland General Hospital (MGH), the following systems need to be commissioned: a new 2000KVA normal power substation, a new 500 KW emergency generator and paralleling switchgear, three new automatic transfer switches and distributions, 2 new 650 ton electric centrifugal chillers and 650 ton cooling towers, temperature and humidity systems, and duct work, air handlers, dampers, and fans [21]. There has to be an easily accessed platform for data exchange and integration. BIM is envisaged to overcome the problem of interoperability and provide easy access for these massive data. When scoping which facilities need to be commissioned, similar projects' information could be retrieved from BIM database, as reference for the decision. The graphical interface of BIM could also improve the collaboration among owner, designer, and contractor. Schedules and commission facilities could be predesigned and stored in BIM models by each facility/zone/room.

These plans are shown as timeline, which act as a simulation of actual commissioning practice. Thus, the logical faults and collision between activities can be easily identified. Moreover, design errors and conflicts of plumbing, HVAC, and electrical from different team could be discovered in the integrated view of BIM. In the design phase, different commissioning tasks have to be assigned to the specified experts for individual commissioning. After all these commissioning subtasks are approved, the whole system has to be commissioned together. For example, plumbing and electrical systems in a room need to be commissioned in a designed order. After both are approved, it must be ensured, these systems could work together successfully. Thus, the overall commissioning task must be done. When commissioning the plumbing or electrical system in different areas, a logical order must be specified. Simulating this schedule and identifying an optimal alternative will reduce the commissioning cost and time. Figure 6 depicts the BIM-based commissioning streamline.

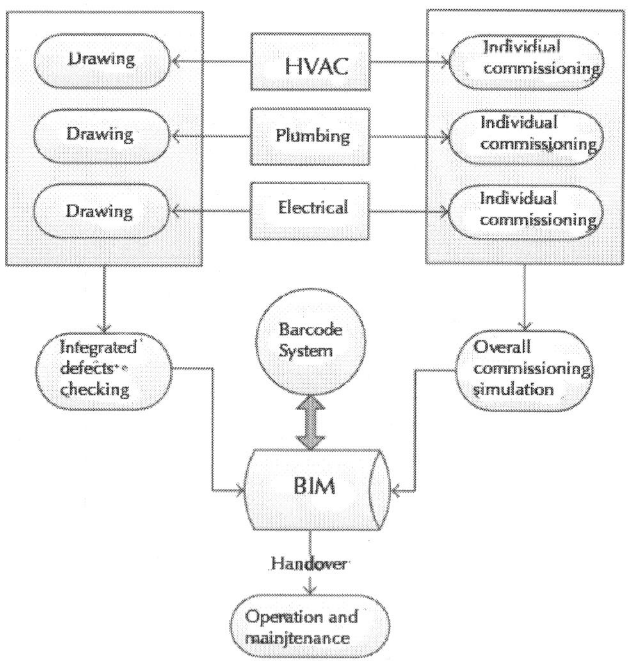


**Figure 6:** BIM-based commissioning streamline.

Barcode system could also be incorporated into BIM for the ease of accessibility of commissioning documents. Each commissioning facility is assigned a unique ID in the BIM model for storing the properties and relevant documents when it is designed. Each ID could be associated with one barcode. The commissioning team could scan the barcode and retrieve the product data, operation data, and maintenance manuals right in the field. Commissioning critical tasks and checklists could be pushed on to a mobile device. The results will be automatically uploaded to BIM central database in a standard format after commissioning tasks.

As project proceeds, these data are handed over to the operational phase as well as the updated BIM model. Besides correcting design defects, simulating commissioning schedules, and bringing easy access for information, bringing commissioning to design phase through BIM could also improve energy saving performance in operation and maintenance phase.

# CASE STUDY

The project of Shanghai Disaster Tolerance Center is an ideal example of bringing FM to design stages through BIM. Travelling path of facility manager is predesigned in BIM thus reducing the maintenance time and providing easy access to the location information of facilities.

## Project Overview

Shanghai Disaster Tolerance Center is in the north of the North Industrial Park in Shanghai, China. It was designed in September of 2010 as a State Grid Corporation of centralized information systems data center. The construction area is 28,124 square meters with one underground layer and four floors on the ground. The diesel generator room and pump are 9.1 meters in height, with construction area of 1,703 square meters. Shanghai Municipal Electric Power Company is the construction company. Shanghai Modern Architectural Design Co., Ltd. is the design company. This project is complicated in facility systems with a tight schedule. High-standard requirements of materials and labor cost control are other characteristics of this project. BIM has been decided to be the tool to bring the FM work into design stage for predesign and simulating the maintenance work in FM.

## BIM Services Content

Accurate BIM model of the mechanical, electrical, construction, and interior decoration are created based on 2D drawings provided by the owner. Clash detection of pipelines is conducted and optimized. BIM model is used for scheduling and guiding the on-site construction work. A 4D construction simulation is also conducted based on the BIM model. Security control and quantity takeoff are based on analysis of pedestrian stream. Construction schedule needs to be incorporated into model in order to visualize the construction process in BIM model. Last but not least, a database platform is developed to read BA surveillance data and conduct real time positioning in 3D mode to get location of facilities thus facilitating maintenance work. This improves the monitoring ability and security level of the disaster tolerance center.

## BIM-Based Travelling Path Optimization in Maintenance

During the process of maintenance, FM personnel have to identify the components' location, getting access to the relevant documents, and finally, the maintenance information. Location information of facility could help facility managers efficiently identify the location of specific building components, especially for those who outsource the FM tasks. Conventionally, they have to log on to different electronic document management systems (EDMSs) and toggle between multiple databases to get the location information, relevant maintenance manuals, and warranty documents. BIM could integrate all these information together in a graphical view. By predesigning the travelling path in the maintenance job, travelling time is well scheduled and reduced and latent hazards could be avoided. Traditionally, after identifying the building number and room number, FM personnel just go to the maintenance spot through a normal path, which may be not the shortest path. Moreover, latent hazards are not identified because of lack in relevant knowledge. Figure 7 depicts the normal travelling path and latent hazards. Since there are different kinds of latent hazards in different areas, the travelling path needs to be identified with the knowledge of all departments of FM teams. After discussion between spatial experts and FM team, an optimal path is specified and incorporated into BIM database, which is safe and consumes the littlest time.

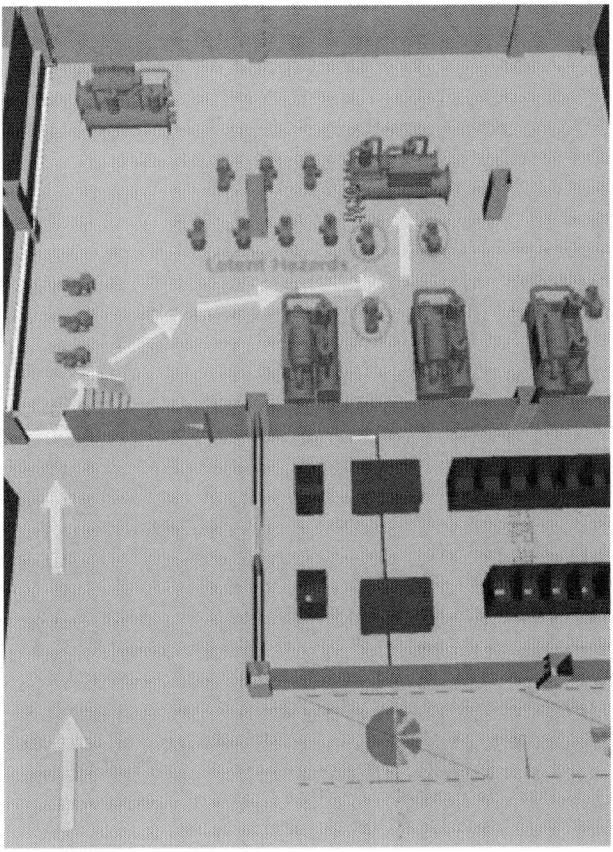

**Figure 7:** Traditional travelling path and latent hazards.

In the following scenario, using BIM to design, optimize, and simulate the path of troubleshooting, the reciprocating compressor is illustrated. Figure 8 depicts the reciprocating compressor with problem.

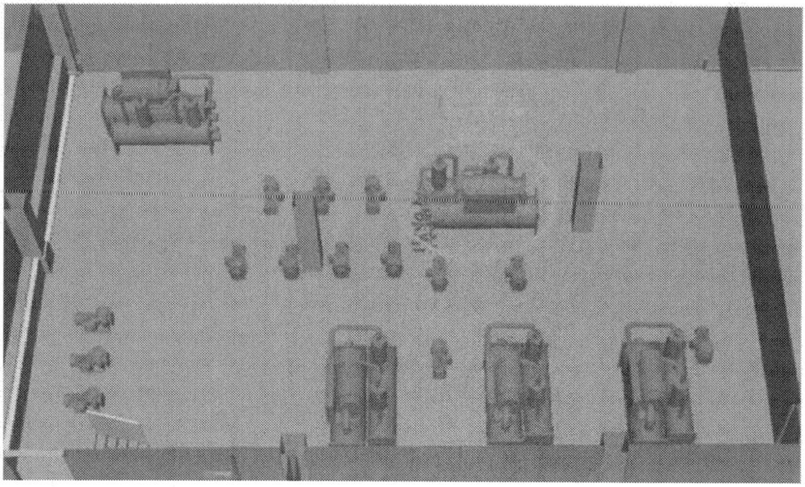

**Figure 8:** Targeted reciprocating compressor.

Firstly, FM staff receives a manual request of troubleshooting the reciprocating compressor with ID 98241620-001f-49d7-94e9-7104b0a3a93d in the underground floor of the Shanghai Disaster Tolerance Center. He then opens the BIM model and searches with this ID. A section view of Figure 8 is shown, and the targeted reciprocating compressor is highlighted. Access to the maintenance manuals, warranty documents, and maintenance history is also provided in the BIM model. After choosing the action of "go the maintenance site," an optimal path is visualized which is safe and timesaving as depicted in Figure 9. A third person view is also provided for the simulation of travelling. Arrow keys in keyboard can be used to control the character. Figure10 depicts the third person view. The FM staff follows the path and troubleshoots the reciprocating compressor with maintenance manuals in a mobile device. Reports are uploaded to BIM central database as history. Status of the reciprocating compressor is updated as "Normal."

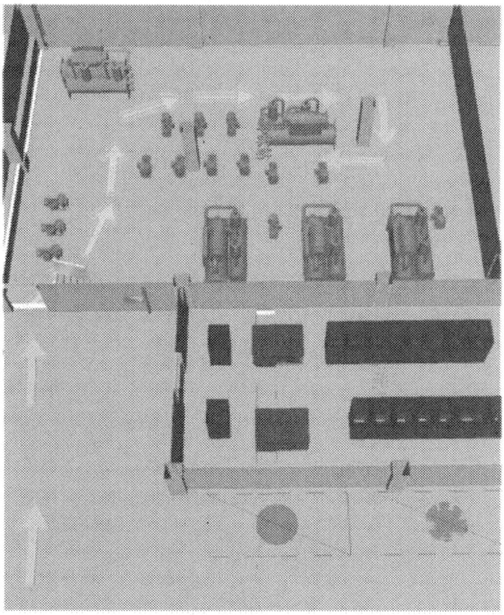

**Figure 9:** Optimal maintenance travelling path.

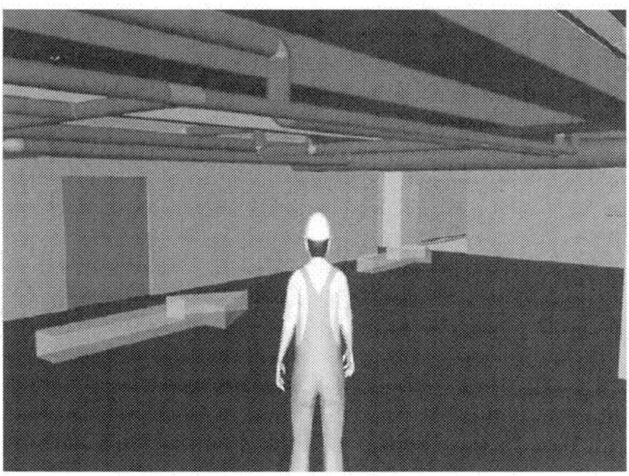

**Figure 10:** Third person view of travelling.

In this case, BIM is utilized as a database and visualization platform to predesign the travelling path in the maintenance job. With knowledge of FM experts, travelling time is well scheduled and reduced, and latent hazards are avoided in the design stage.

# CONCLUSION AND FUTURE WORK

This study developed a framework of considering FM in design stage with BIM. The contribution of this paper is the development of an innovative framework, which integrates FM work into early design stage via BIM. Furthermore, one aspect of the whole framework is validated for the proof of the concept. An innovational concept of gathering designers with the FM team through BIM is proposed for strengthening collaboration as well as information sharing and gathering. The purpose is to avoid and reduce the potential issues such as rework and inappropriate allocation of workspace in the operational phase. As little research has identified the approach and benefit of integrating FM with early design stage, this study aims at bridging this gap by providing a working pattern of providing the essential information with BIM. Due to the difficulty of altering the main structure and core service areas in the operational phase, it is practical to design for adaptability by considering operational condition and the facilities' own attributes. It is very difficult to achieve without the relevant information from FM team and appropriate integration platform. With the ease of access to lifecycle information of all the building components BIM provided, the proposed building plan could be optimized, and lifecycle cost could be reduced with the FM knowledge and experience.

# ACKNOWLEDGMENTS

CCDI of China provided support for the traveling path optimization case study, which is a large architectural consulting firm that provides integrated services for urban construction and development. Data and projects of Shanghai Modern Architectural Design Co., Ltd., Shanghai Municipal Electric Power Company, and Grid Corporation of centralized information systems data center belong to the CCDI company. The contents of this study reflect the views of the authors who are responsible for the facts and the accuracy of the data presented herein. The contents do not necessarily reflect the official views or policies of these companies.

# REFERENCES

1. International Facility Management Association.
2. B. Becerik-Gerber, F. Jazizadeh, N. Li, and G. Calis, "Application areas and data requirements for BIM-enabled facilities management," Journal of Construction Engineering and Management, vol. 138, no. 3, pp. 431–442, 2012.
3. B. Nutt and P. McLennan, Facility Management: Risks and Opportunities, Blackwell Science, 2000.
4. D. G. Cotts, K. O. Roper, and R. P. Payant, The Facility Management Handbook, Amacom Books, 2010.
5. E. Erdener, "Linking programming and design with facilities management," Journal of Performance of Constructed Facilities, vol. 17, no. 1, pp. 4–8, 2003.
6. R. Sacks, I. Kaner, C. M. Eastman, and Y. S. Jeong, "The Rosewood experiment—building information modeling and interoperability for architectural precast facades," Automation in Construction, vol. 19, no. 4, pp. 419–432, 2010.
7. V. Singh, N. Gu, and X. Wang, "A theoretical framework of a BIM-based multi-disciplinary collaboration platform," Automation in Construction, vol. 20, no. 2, pp. 134–144, 2011.
8. J. P. Duarte, G. Celani, R. Pupo, et al., "Inserting computational technologies in architectural curricula," in Computational Design Methods and Technologies: Applications in CAD, CAM and CAE Education, N. Gu and X. Wang, Eds., IGI Global, Hershey, Pa, USA, 2010.
9. X. Wang and P. S. Dunston, "Comparative effectiveness of mixed reality-based virtual environments in collaborative design," IEEE Transactions on Systems, Man and Cybernetics C, vol. 41, no. 3, pp. 284–296, 2011.
10. M. R. Devetakovic and M. Radojevic, "Facility Mangement: a paradigm for expanding the scope of architectural practice,"

International Journal of Architectural Research, vol. 1, no. 3, pp. 127–139, 2007.

11. International home of openBIM. 1994,
12. D. Arditi and M. Nawakorawit, "Designing buildings for maintenance: designers' perspective," Journal of Architectural Engineering, vol. 5, no. 4, pp. 107–116, 1999.
13. H. Kim, E. Jenicek, and A. Stumpf, "Early design energy analysis using bims (building information models)," in Proceedings of the Construction Research Congress, pp. 426–436, April 2009.
14. ASHE, Healthcare Facility Commissioning Guideline, ASHE, Chicago, Ill, USA, 2010.
15. J. W. Korka, A. A. Oloufa, and H. R. Thomas, "Facilities computerized maintenance management systems," Journal of Architectural Engineering, vol. 3, no. 3, pp. 118–123, 1997.
16. U. S. G. B. Concil, Green Building Facts, 2009.
17. P. A. Torcellini, S. J. Hayter, and R. Judkoff, "Low-energy building design—the process and a case study," in Proceedings of the ASHRAE Annual Meeting, pp. 802–810, June 1999.
18. B. Dong, K. P. Lam, Y. C. Huang, and G. M. Dobbs, "A comparative study of the IFC and gbXML informational infrastructures for data exchange in computational design support environments," in Proceedings of the Building Simulation, 2007.
19. US Environmental Protection Agency,
20. DOE, BUILDING COMMISSIONING: The Key to Quality Assurance, Rebuild America Guide Series, DOE, Washington, DC, USA, 1998.
21. G. C. Lasker, H. Y. Dib, and C. Chen, "Benefits of implementing building information modeling for healthcare facility commissioning," in Computing in Civil Engineering, vol. 2011, pp. 578–585, 2011.

Chapter 6

# Developing Mobile- And Bim-Based Integrated Visual Facility Maintenance Management System

Yu-Cheng Lin and Yu-Chih Su

Department of Civil Engineering, National Taipei University of Technology, No. 1 Chung-Hsiao E. Road, Section 3, Taipei 10608, Taiwan

## ABSTRACT

Facility maintenance management (FMM) has become an important topic for research on the operation phase of the construction life cycle. Managing FMM effectively is extremely difficult owing to various factors and environments. One of the difficulties is the performance of 2D graphics when depicting maintenance service. Building information modeling (BIM) uses precise geometry and relevant data to support the maintenance service of facilities depicted in 3D object-oriented CAD. This paper proposes a new and practical methodology with application to FMM using BIM technology. Using BIM technology, this study proposes a BIM-based facility maintenance management (BIMFMM) system for maintenance staff in the operation and maintenance phase. The BIMFMM system is then applied in selected case study of a commercial building project in Taiwan to verify the proposed methodology and demonstrate its effectiveness in FMM practice. Using the BIMFMM system, maintenance staff can access and review 3D BIM models for updating related maintenance records in a digital format. Moreover, this study presents a generic system architecture and its implementation. The combined results demonstrate that a BIMFMM-like system can be an effective visual FMM tool.

# INTRODUCTION

Facility maintenance management (FMM) in the operation phase of facility's life cycle has become an important topic for research and academic study. Managing maintenance information about facilities contributes to successful facility management (FM). Managing FMM work effectively can be extremely difficult on the operation phase owing to various types of equipment and facilities. Furthermore, it is inconvenient for maintenance staff to maintain those facilities by relying on paper-based documents. The latest information technology solutions provide improved FMM. Unlike the manufacturing industry, information technology is limited in its use and application in construction [1], and most of the management work is done by human labor, which is inefficient and sometimes error-prone.

Regarding FMM, maintenance staff usually refers to information such as specifications, checklists, maintenance reports, and maintenance records. Maintenance staff must record maintenance results on hard copies. Consequently, there can be significant gaps in data capture and entry. Such means of communicating information between the facility location and the management office are ineffective and inconvenient. According to the survey findings regarding maintenance work on a commercial building in Taiwan, the primary problems regarding data capture and sharing during the FMM process are as follows: (1) the efficiency and quality are low, especially through document-based media, (2) it is not easy to reference relevant detailed information on facilities, (3) there are data reentry problems, and (4) the use of desktops for operating BIM models cannot be extended to maintenance management service at facility location effectively. However, few suitable platforms exist to assist maintenance staff in using integrated FMM information system from BIM models and sharing maintenance information directly at the facility's location.

The performance of FMM can be enhanced by using web technology for information sharing and communication. In this study, the FMM work includes inspection and maintenance works. Building Information Modeling (BIM) uses precise geometry and relevant data to support the maintenance service of facilities depicted in 3D object-oriented CAD. By integrating web and BIM technologies, the effectiveness of FMM work is enhanced and improved (see Figure 1). In order to enhance the effectiveness of FMM work on commercial buildings, this study presents a novel system called BIM-based Facility Maintenance Management (BIMFMM) system for the acquisition and tracking of maintenance information and provides an information sharing platform for maintenance staff using a webcam-enabled notebook or tablet. Integrating the web and BIM technologies, information and data entry mechanisms can help to improve the effectiveness and convenience of information flow in the FMM process. The primary objectives of this study include (1) applying BIM and web technologies to increase the efficiency of collecting maintenance data and information, (2) accessing web technologies directly to link detailed information with BIM models of facilities, and (3) exploring the limitations of the system, addressing problems, and providing suggestions based on the implementation of the pilot case study. The BIMFMM system is applied to a commercial building in Taiwan to verify our proposed methodology and demonstrate the effectiveness of the FMM process. The combined results demonstrate that the BIMMM system can be a useful BIM-based FMM platform by utilizing web and BIM technologies.

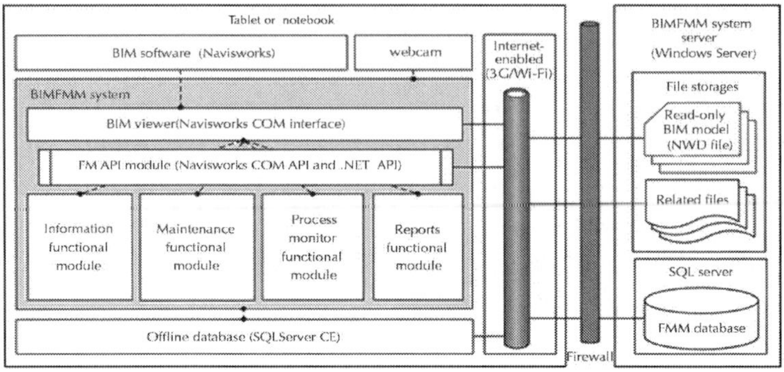


**Figure 1:** Overview of the BIMFMM system framework.

## Related Research Studies

BIM digitally contains precise geometry and relevant data needed to support the design, procurement, fabrication, and construction activities to describe 3D object-oriented CAD [2]. BIM is a digital tool that supports continual updating and sharing of project design information [3]. BIM is the process of generating and managing building data during a building life cycle [4]. BIM technology has the potential to enable fundamental changes in project delivery to support a more integrated, efficient process [5]. Much previous research has examined BIM issues in construction. There are many core benefits, barriers, frameworks, and recommendations on BIM usage cited in previous work on supporting decisions and improving processes throughout the lifecycle of a project [2, 6–14]. Related to the design phase of a project, these topics include parametric modeling, BIM at different levels of development (LOD), identification of design conflicts and analysis, green design, design simulation, cost estimation, and accurate geometric representation of all facilities [2, 7, 8, 15–27]. During the construction phase, these benefits using BIM in construction include less rework, reduction in requests for information and change orders, customer satisfaction through visualization, improved productivity in phasing and scheduling, faster and more effective construction management with easier information exchange, accurate cost estimation, effective supply chain, and visualizing safety analysis [2,

28–35]. During the operation phase, these benefits include control of maintenance management process, integrated life cycle data, rapid and accurate information about update and change activities, and more effective FM with easier information exchange [2, 5, 8, 17, 18, 29, 36–42].

The BIM approach, which is used to retain facility information in a digital format, facilitates easy updates of FMM information in a 3D CAD environment. Although there were many practical applications for using BIM in the maintenance management stage, one of the challenges in broader application of BIM models to FMM is that currently the use of PC desktops limits on-site use of BIM models during the maintenance and inspection process. Another problem for most facilities is that vertical position can be difficult to illustrate clearly based on traditional 2D drawings. With the use of mobile devices executing the BIM models, these BIM models need to be processed in advance and reduced to a smaller file size to be used in mobile devices. This study will explore and make recommendations to solve these problems. In order to assist maintenance staff in obtaining the corresponding BIM model automatically for FMM, this study develops a proposed system to integrate web technology to automatically connect the BIM models. This study enhances FMM service using web technology integrated with the BIM approach. By using the web technology, users can quickly click the corresponding BIM model of a facility and access basic information and maintenance problems while managing FMM information during the operation phase.

# SYSTEM SCHEMATIC DESIGN

The application of BIM technology in the FMM both inside and outside of the buildings supports maintenance staff in handling FMM via the 3D BIM models. By accessing the mobile device, maintenance staff can obtain the corresponding BIM model of the facility and directly access FMM information about the facility such as instruction manuals, photos, video of operations, maintenance history, and manufacturer information. Furthermore, a 3D BIM model improves upon traditional 2D drawings that can make it difficult to illustrate the vertical location or position of facilities.

The BIMFMM system consists of subsystems for BIM, mobile devices, and a hub center. The mobile devices subsystem is located on the client side, while the BIM models and hub center subsystem are on the server side. Each subsystem is briefly described below.

## BIM Modules Subsystem of the BIMFMM System

In this study, BIM is used as an information model in the BIMFMM system. The BIM is applied to capture and store information about the facility, including basic descriptions, parameter-related information, maintenance records, and FMM reports. Autodesk Revit Architecture and Revit MEP were used to create and maintain the BIM model files. Autodesk Navisworks was used to integrate and read the BIM models of facilities. Information integration with the 3D BIM models was achieved using the Autodesk Navisworks API and Microsoft Visual Basic.NET (VB.NET) programming language. The BIMFMM system was developed by integrating the 3D BIM models of facilities and maintenance-related information using Navisworks API programming. Open Database Connectivity (ODBC) was utilized to integrate acquired data from different software programs and all maintenance information, such that BIM files can be exported to an ODBC database for connection with the BIMFMM system.

### *Mobile Devices Subsystem of the BIMFMM System*

There are two mobile devices used in the BIMFMM system. An Acer Iconia W700P tablet is used as the webcam-enabled tablet hardware. The Acer Iconia W700P tablet runs on Windows 8. An HP Pavilion notebook is used as the webcam-enabled notebook hardware. The HP Pavilion notebook runs the Windows 7 operating system. All data in the tablet and notebook are transmitted between the client and the server sides directly through the web via Wi-Fi or 3G.

## Hub Center Subsystem of the BIMFMM System

The hub center is an information center in the BIMFMM system that enables all participants to log onto a hub center and immediately obtain information required for FMM. Users can access different information and services via a single front-end access point on the Internet. For example, maintenance staff can log onto the hub center and securely access the latest FMM schedule information. FM managers can check maintenance status, results, and various other inspection-related data. All facilities-related pieces of information acquired within the hub center subsystem are recorded in a centralized system database. Maintenance staff can access required information via the hub center subsystem based on their access privileges.

The amount of maintenance information stored will increase over time if all FMM pieces of information are recorded in the BIM model. Because BIM models cover a wealth of building information, central BIM models storage space should be reserved for crucial information, such as spatial information, facility ID and name of facility, facility location, and other critical information. In order to keep the system performance at an acceptable level, the information derived by other applications should be stored in an external location. Therefore, there is one database designed in the BIMFMM system, called the FMM database. The central BIM models stores only basic information (such as position, ID and name of facility, and key parameter information of components). Related maintenance data and information are stored in the FMM database.

The accuracy of the central BIM models will directly affect FMM operations in the BIMFMM system. In order to prevent too many users from simultaneously using central BIM models and in turn affecting the accuracy of the BIM models, the BIM engineer can update central BIM models and export into read-only BIM models (NWD file) directly in the server side. The latest read-only BIM models on the client side automatically re-syncs when read-only BIM models change on the server side. In this framework, all building facility pieces of information from BIM can be saved and updated in the read-only BIM models without accessing the central BIM models directly. Furthermore, central BIM models and read-only BIM model are saved on the server side. Figure 1 shows overview of the BIMFMM system framework.

In the BIMFMM system, major three roles are involved in FMM including BIM engineer, FM manager, and maintenance staff. To ensure that the FMM operation does not affect the maintenance operation of the central BIM models, this study utilizes client-server system architecture. In the BIMFMM system, the read-only BIM model stores all facility basic information and location in the server side. Also, only BIM engineers are allowed to access and edit central BIM models by using BIM software and export into the read-only BIM models directly on the server side. On the client side, the FM manager and maintenance staff reference facility information through the read-only BIM models and edit FMM information through the FMM database in the BIMFMM system.

The BIMFMM system server supports four distinct layers, each with its own responsibilities: management, data access, application, and presentation (see Figure 2). This following section describes the distinct layers in the BIMFMM system.

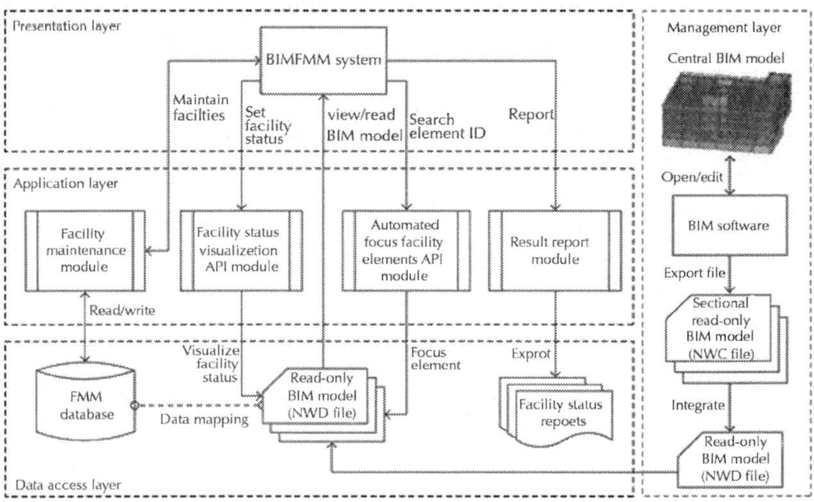


**Figure 2:** System and module framework of the BIMFMM system.

The management layer provides BIM engineers with tools to edit and manage central BIM models by using BIM software. BIM engineers can create and integrate the read-only BIM models saved in the server through the Internet.

Regarding the data access layer of the BIMFMM system, the FMM database stores all facilities maintenance records, while the read-only BIM models store complete facility information including facility number, name, and type in the BIM models. The FMM database records detailed maintenance information in accordance with the facility ID. The primary key establishes a relationship between facility ID and the main index. Therefore, information can be used for data association for data mapping to retrieve complete facilities maintenance information based on facility ID between read-only BIM models and FMM database.

The application layer defines various applications for major system and API modules. These applications offer indexing, BIM model data updates and transfers, facility status visualization, and report generation functions. The application layer integrates and uses BIM software to open the BIM models by using developed API modules. Finally, the application layer can automatically acquire data and analyze BIM models based on a request and then send the results back to the client side.

The presentation layer is the main implementation platform of the BIMFMM system. During the FMM process, the FM manager and maintenance staff can use a tablet (client side) and utilities in the BIMFMM system for the FMM operation. The presentation layer displays the location information of BIM model automatically, records maintenance information, illustrates different conditions and status of FMM, queries the history, and exports reports on FMM results.

# SYSTEM DEVELOPMENT

The BIMFMM system server is based on the Microsoft Windows Server 2008 operating system with an SQL Server 2008 R2 as the database. The BIMFMM system is developed using VB.NET programming, which is easily incorporated with ADO.NET to transact FMM and BIM information with an SQL Server database. The BIMFMM system consists of three different user areas, maintenance staff, FM manager, and BIM engineer areas. Access to the BIMFMM system is password-controlled.

## System Functionality Description

This section describes the implementation of each major functionality module in the BIMFMM system.

### *FMM Information Functional Module*

The functional module provides maintenance staff with detailed FMM information on facilities by reviewing 3D BIM models. This module enables all maintenance staff to refer to related FMM information and historical maintenance records for the selected facility quickly and easily in the BIM-based environment. This module allows maintenance staff to refer to basic information and specifications associated with BIM models during the FMM process. This module also has a search function that enables the information to be found and retrieved easily.

### *Maintenance Functional Module*

Maintenance staff can download up-to-date maintenance records through the BIM models and enter facility maintenance results directly into the BIM models. Additionally, the module can automatically produce the corresponding maintenance forms through the BIM models. Tablets display the checklist for every facility maintenance task. Maintenance staff can record maintenance information such dates, conditions, inspection results, descriptions of problems that have arisen during maintenance, and recommendations. Furthermore, maintenance staff can also check tasks that do not pass the inspection and select relevant tasks from lists in the BIM models. One of the benefits of the module is that maintenance results and records can be transferred between a tablet and the BIMFMM system by real-time synchronization, eliminating the need to enter the same data more than once.

### *Process Monitor Functional Module*

This functional module is designed to enable FM managers to monitor the FMM process. The process monitor module provides an easily accessed and portable environment where maintenance staff can trace and record all maintenance information and statuses through the visualized and colorized BIM model.

## Reports Functional Module

Users can easily access the FMM reports functional module to identify needs and analyze FMM results information. Authorized records for interfaces can be extracted and summarized for the final FMM result-related reports. Furthermore, all FMM reports can be extracted using commercially available software such as Microsoft Excel.

## System API Modules Description

In order to integrate the system with the BIM models, the following API modules are developed in the BIMFMM system.

## Automated Focus Facility Elements API Module

This module allows users easily and quickly to access the related BIM models by entering ID code attached in surface of the facility. When the user enters ID code, this module will automatically identify the facility angle and facility location in the corresponding BIM models automatically for FMM.

## Facility Status Visualization API Module

The module provides the visualization functionality for FMM status through a visualized BIM model. Through a systematic FMM analysis of test results, the module displays different colors to illustrate various conditions and FMM status (such as qualified inspection, required repair status, and obsolete facility). Users can access the overall different maintenance conditions and FMM statuses quickly through the visualized BIM model.

There are two subsystems in the BIMFMM system. The first subsystem is the API monitoring subsystem for BIM engineers located on the server side. This subsystem deals with integration services of BIM models in the BIMFMM system. These services include updating facility maintenance information. Another subsystem is the maintenance subsystem located on the client side. This maintenance subsystem is developed for maintenance staff and FM managers to deal with FMM operations in the facility's location, such as clicking the BIM model of the facility, recording FMM, and reporting FMM results.

## System Process Description

There are three processes used in the BIMFMM system including the system initialization process, FMM information monitoring process, and maintenance implementation process.

### System Initialization Process

The purpose of the system initialization process is to provide adequate information on FMM operations. The system initialization process includes BIM models initialization and facility information initialization.

*BIM Models Initialization:* The BIM model must provide all pieces of information and related models on a facility as an information requirement for facility maintenance operations. When the BIM models save complete facility information, the BIM engineer needs only to use BIM software (such as Revit) to create BIM models first. After the BIM models creation by Revit can export into many read-only BIM models (NWC file). All exported NWC files will be integrated a single read-only BIM model (NWD file) using Navisworks. Finally, the read-only BIM models (NWD files) can be downloaded on the client side of the BIMFMM system for FMM usage. When central BIM models change, BIM engineers only need to update central BIM models and export to read-only BIM models (NWD files); the BIMFMM system will automatically update read-only BIM model on the client side of the BIMFMM system.

*Facility Information Initialization:* Facility information initialization is an important task in the FMM. After BIM models initialization, FMM staff adds new facility information and electronic documents directly in the system. The information will be saved in the FMM database. Furthermore, the facility information must be linked and associated with BIM models to create the relationship of facility information and BIM models in the MM database.

After the facility information initialization process is completed, FM manager and maintenance staff may utilize BIMFMM system to handle the following FMM information monitoring and maintenance implementation tasks.

### *FMM Information Monitoring Process*

When the system initialization process is completed, FM manager can view and access BIM models (NWD file) in the BIMFMM system and then open the BIMFMM system monitoring API module. When maintenance staff handles the FMM process, FM managers can monitor and refer the newest state of FMM with different color visualization of BIM models through the BIMFMM system.

## Maintenance Implementation Process

During the maintenance implementation process, the maintenance list varies according to the maintenance task categories. The design lets maintenance staff work on maintenance operations effectively according to the task categories and maintenance list. Maintenance staff can utilize the Web-enabled Tablet PC to access the BIMFMM system and show all the task categories and maintenance list based on different levels of access. After maintenance staff selects a particular task category, the system shows the history task form for that category. Maintenance staff can view the other task form, edit the unfinished task form, or add a new task form. When maintenance staff selects or adds a task form, the system retrieves facility information from the read-only BIM model based on the task types. Furthermore, a list of all related maintenance and results will be illustrated with the BIM model for FMM work preparation. Maintenance staff can access inspection information and the maintenance status effectively. During the maintenance implementation process, maintenance staff can use the system directly and select the corresponding BIM model of the facility. When the system receives the facility ID, the system automatically displays the facility's basic information and historical maintenance data in the BIM model. Furthermore, the facility's BIM model will be selected, focused, and highlighted using different color. User can obtain basic information on the facility by clicking the BIM model, or selecting from a maintenance list. After selecting the facility through one of the three methods, the maintenance staff can handle maintenance work and record the status and result of maintenance. Finally, all maintenance records and pieces of information are stored in the FMM database.

During the process of maintenance operations, the maintenance status also can be enhanced by color visualization in the BIM model through the facility status visualization API module. Through the functionality that visually depicts the status of maintenance list items, the BIMFMM system will get related maintenance status from the maintenance list in the FMM database. Furthermore, the BIM model will visualize different colors based on the each maintenance status in the facilities, and other elements in BIM model will be displayed in translucent white to enhance the visualization effect (see Figure 3).

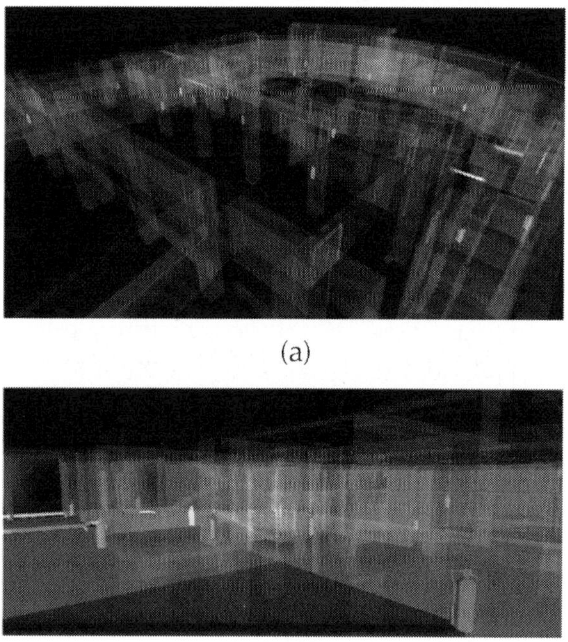

(a)

(b)

**Figure 3:** The different maintenance statuses of FMM through visualized and colorized BIM model.

Integrated with the above design concept, more complex operating procedures of FMM are simplified and developed in the BIMFMM system. One of the major characteristics of the BIMFMM system is to provide users an easy-to-use visualization for handling FMM work. By clicking the list, each task form will show the list of facilities requiring maintenance, historical maintenance information, and the status and condition of facilities maintenance. By clicking the corresponding BIM model of the facility, the BIM model are linked and illustrated quickly and effectively in facility location. All maintenance results are sent back and saved in the main BIM model. The proposed approach provides a means to update the facility information of the BIM model and FMM information synchronization. Finally, in order to let FMM engineers apply the system easily and effectively, the layout of the system is designed based on FMM engineers' suggestions. Figure 4illustrates the system process flowchart used in the BIMFMM system. Figure 5 shows the graphical user interface (GUI) of the BIMFMM system.

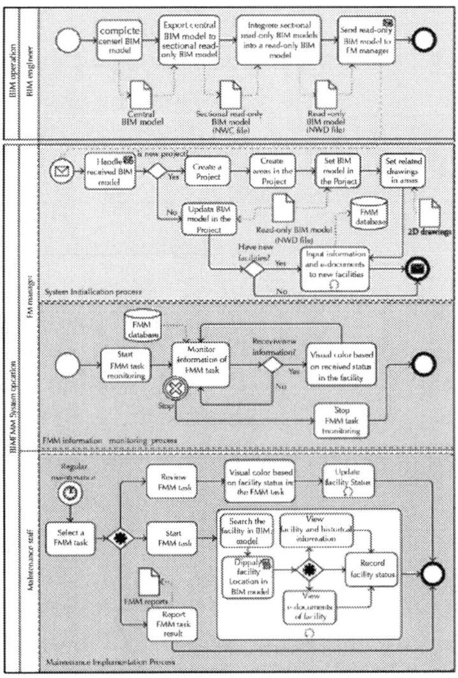


**Figure 4:** The process diagram used in the BIMFMM system.

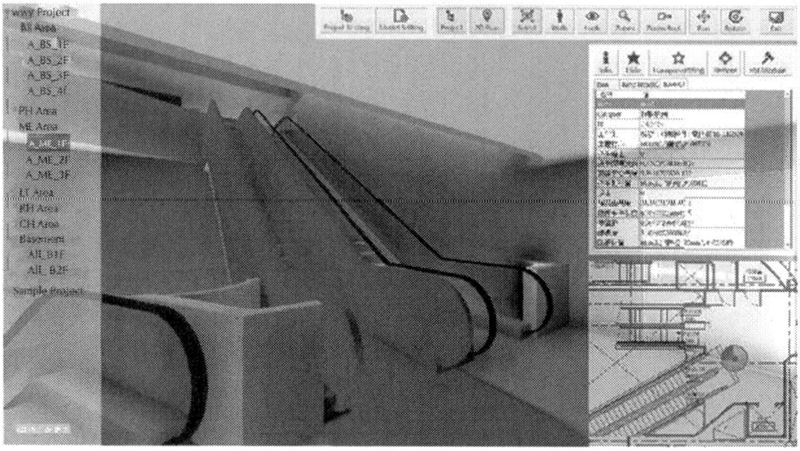

**Figure 5:** GUI of the BIMFMM system.

# SYSTEM VALIDATION

## Pilot Case Study

This study is applied to a building in Taiwan for the case study. This study utilizes a BIMFMM system in the FMM for the building. Existing approaches for tracking and managing FMM work rely on paper-based records. The bulk of FMM work was paper-based and documented by repeated manual entry, although an FMM system was developed for a standalone software application. Therefore, maintenance staff in the FM division utilized the BIMFMM system to enhance FMM work in the pilot case study.

After the critical facilities were selected for FMM work, the unique ID for each facility was entered into the BIMFMM system database for quick search. Before the FMM work began, the maintenance staff could check the facility list from webcam-enabled tablets, refer to the relevant information, and begin preparation work without printing any paper documents. During the FMM process, the maintenance staff selected the relevant BIM model. The BIMFMM system showed the basic information and BIM model of the facility. Maintenance staff could then check further detailed information like maintenance instructions, notifications, and accessories list, all of which are supported by BIMFMM. After the FMM work, maintenance staff entered the results of maintenance, edited the description in the tablet, and provided the updated information to the system. When a facility required repairs, the system also provided the manufacturer's problem information for immediate reference. Finally, the facilities manager and the authorized maintenance staff accessed the updated information simultaneously from their offices. Figure 6 illustrates maintenance staff using webcam-enabled tablet for FMM work in the pilot case study.

**Figure 6:** Staff using webcam-enabled tablet for FMM work in the pilot case.

# EVALUATION AND RESULTS

Overall, the field test results indicate that the application of BIM is an effective tool for FMM in a building. All BIM models survived use in the pilot test over the two-month testing period. Approximately 18 users participated in field trials of the FMM process. The BIMFMM system was installed on the main server in the FM division of the building.

During the field trials, verification and validation tests were performed to evaluate the system. Verification aims to evaluate whether the system operates correctly according to the design and specification; validation assesses the usefulness of the system. The verification test was carried out by checking whether the BIMFMM system could perform tasks as specified in the system analysis and design. The validation test was undertaken by asking selected case participants to use the system and provide feedback by answering a questionnaire. There were 25 participants involved in the evaluation test. To evaluate system function and the level of satisfaction with the system's capabilities, the users of the system were asked to grade the conditions of system testing, system function, and system capability separately, compared with the typical paper-based FMM approach. Some comments for future improvements to the BIMFMM system were also obtained from the case participants through the user satisfaction survey. Finally, Table 1 shows system evaluation result.

**Table 1:** System evaluation result

| System functionality | Mean score |
|---|---|
| Ease of FMM information sharing | 4.4 |
| Reliability | 3.9 |
| Applicability to FMM | 4.7 |
| Use of system | Mean score |
| Ease of use | 4.4 |
| User interface | 4.4 |
| Overall system usefulness | 4.4 |
| System capability | Mean score |
| Reduces unnecessary time | 4.3 |
| Ease of finding maintenance information | 4.4 |
| Improves maintenance problem tracking | 4.0 |
| Enhances visual maintenance management | 4.4 |
| Enhances maintenance problems illustration | 3.9 |

Note: the mean score is calculated from respondents' feedback on five scale questionnaire: 1 (strongly disagree), 2, 3, 4, and 5 (strongly agree).

The 92% of users obtained from user satisfaction survey indicates that the BIMFMM system is quite adaptable to current FMM practices in a building and is attractive to users. The overall result implies that the BIMFMM system is considered to be well designed and could enhance current time-consuming FMM processes. The over 98% satisfaction rate also indicates that the visual BIM model providing FMM support is very helpful. The 98% satisfaction rate for the BIMFMM system directly accessing the BIM model at facility location is also effective and necessary. Also, no additional work was required to complete documentation beyond the data collection process. The advantages and disadvantages of the BIMFMM system identified from the pilot study are identified.

# LIMITATIONS AND BARRIERS

The findings of this case study revealed several limitations of the BIMFMM system. The following are inherent problems recognized during the case study.

1. It was difficult for new users to operate BIM model in the BIMFMM system. Some maintenance staff was initially unfamiliar with BIM models. It usually takes time to learn how to use BIM models. In the case study, the use of the BIM system initially lengthened the FMM operation over the traditional approach, since users required time to find the corresponding BIM model and fill out the FMM information in the BIMFMM system. After the user is skilled and familiar with BIM model, the time required by the current approach and the proposed system is almost exactly the same in FMM operations.
2. Based on the case study, BIM engineers needed to keep and update BIM models during the operation phase. When new equipment or facilities are purchased, BIM engineers must build a new BIM model for future maintenance use. Furthermore, the communications between maintenance staff and BIM engineers are necessary and important during the process. Maintenance staff should tell BIM engineers about any problem regarding the

BIM models. The BIM engineers also must notify and discuss with maintenance staff after BIM engineer corrects the BIM models. BIM models require constant maintenance and updates. Another important issue is quality management of BIM models. Although the study proposed that the BIMFMM system help maintenance staff to handle visual facilities maintenance and management work, the advanced management procedures and mechanisms for quality management of BIM models for FMM must be identified and developed in the future. Particularly, the management mechanisms for updating the BIM models should be developed as the next step of BIMFMM system development.

3. Although Navisworks provide user with the ability to access a huge integrated amount of BIM models, the integrated BIM models (NWD file) will become larger than original BIM models. Usually, it will take 2 to 5 minutes to download whole BIM models from the server side when applying BIM models in the tablet for FMM work. Therefore, it is necessary to develop appropriate mechanisms to improve the above problem. For example, the BIM models in the client side will be updated and downloaded only from the server side when BIM models changes in the server.

# CONCLUSIONS

The BIM approach, which is applied to retain facility information in a digital format, facilitates easy updating of FMM information in a BIM environment. Although there were many practical cases for using BIM during the maintenance management stage, the one of facing problem is that typically BIM models could only be used with PC desktops in an office, which limited their use onsite during facility maintenance. However, use of high-end desktops for operating BIM models could not be used effectively by maintenance staff onsite during the maintenance and inspection process. BIM models need to be processed and transferred via smaller files for use with mobile devices, which are more commonly used onsite. In order to assist maintenance staff with obtaining the corresponding BIM model automatically for FMM, this study develops the BIMFMM system to integrate web technology to automatically connect the BIM models. The BIMFMM system not only improves FMM efficiency but also provides a real-time service platform during the FMM process. In the case study, MM staff used webcam-enabled tablets to seamlessly enhance FMM work at facility locations, owing to the system's searching speed and ability to support related information collection and access during the FMM process. Meanwhile, on the server side, the BIMFMM system offers a hub center to provide the FM division with real time monitoring capacity during the FMM process. Integrated with characteristics of 3D BIM model illustration and BIM parametric design, the BIMFMM system quickly shows the necessary maintenance information using a facility's BIM model based on the selected task type and clearly presents the position and height of the selected facility.

In a case study, the application of the BIMFMM system helped to improve the FMM work of a commercial building in Taiwan. Based on experimental results, this study demonstrated that BIM technology has significant potential to enhance FMM work. The integration of BIM technology with web technology helps FM managers and maintenance staff to effectively track and control the whole FMM process. Compared with current approaches, the combined results demonstrate that a BIMFMM system can be a useful mobile BIM-based FMM platform. Based on the case study finding, BIM models must be updated and corrected constantly. Another important issue is quality management of BIM models. The advanced management procedures and mechanisms for quality management of BIM models for FMM needs to be identified and developed in the future. Doing so will be the next step of BIMFMM system development. Finally, the limitations, facing problems, and suggestions are discussed based on the implementation of case studies in this study. Although there are some challenges indicated above, the proposed system has shown a great potential to be used for FMM in building with the promising results shown in this study.

# ACKNOWLEDGMENT

The authors would like to express their appreciation to the staff and managers of D-W Property Company for assistance in providing useful data, valuable information, and helpful comments during system development and test.

# REFERENCES


1. Boddy S, Rezgui Y, Cooper G, Wetherill M. Computer integrated construction: a review and proposals for future direction. *Advances in Engineering Software*. 2007;38(10):677–687.
2. Eastman C, Teicholz P, Sacks R, Liston K. *BIM Handbook: A Guide to Building Information Modeling for Owners, Managers, Designers, Engineers and Contractors*. Hoboken, NJ, USA: Wiley; 2011.
3. Gould F, Joyce N. *Construction Project Management*. 3rd edition. New York, NY, USA: Prentice Hall; 2008.
4. Epstein E. *Implementing Successful Building Information Modeling*. Norwood, Mass, USA: Artech House; 2012.
5. Teicholz P. *BIM for Facility Managers*. Hoboken, NJ, USA: Wiley; 2013.
6. Tse TK, Wong KA, Wong KF. The utilisation of building information models in nD modelling: a study of data interfacing and adoption barriers. *Electronic Journal of Information Technology in Construction*.2005;10:85–110.
7. Shen Z, Issa R. Quantitative evaluation of the BIM-assisted construction detailed cost estimates. *Electronic Journal of Information Technology in Construction*. 2010;15:234–257.
8. Manning R, Messner JI. Case studies in BIM implementation for programming of healthcare facilities.*Electronic Journal of Information Technology in Construction*. 2008;13:446–457.
9. Becerik-Gerber B, Rice S. The perceived value of building information modeling in the U.S. building industry. *Electronic Journal of Information Technology in Construction*. 2010;15:185–201.
10. Jung Y, Joo M. Building information modelling (BIM) framework for practical implementation.*Automation in Construction*. 2011;20(2):126–133.
11. Barlish K, Sullivan K. How to measure the benefits of BIM—a case study approach. *Automation in Construction*. 2012;24:149–159.
12. Bryde D, Broquetas M, Volm JM. The project benefits of Building Information Modelling (BIM)*International Journal of Project Management*. 2013;31(7):971–980.

13. Sebastian R, Van Berlo L. Tool for benchmarking BIM performance of design, engineering and construction firms in the Netherlands. *Architectural Engineering and Design Management*. 2010;6:254–263.
14. Becerik-Gerber B, Kensek K. Building information modeling in architecture, engineering, and construction: emerging research directions and trends. *Journal of Professional Issues in Engineering Education and Practice*. 2010;136(3):139–147.
15. Li G, Ding L, Wang J. Construction project control in virtual reality: a case study. *Journal of Applied Sciences*. 2006;6(13):2724–2732.
16. Leite F, Akcamete A, Akinci B, Atasoy G, Kiziltas S. Analysis of modeling effort and impact of different levels of detail in building information models. *Automation in Construction*. 2011;20(5):601–609.
17. Staub-French S, Khanzode A. 3D and 4D modeling for design and construction coordination: issues and lessons learned. *Electronic Journal of Information Technology in Construction*. 2007;12:381–407.
18. Fox S, Hietanen J. Interorganizational use of building information models: potential for automational, informational and transformational effects. *Construction Management and Economics*. 2007;25(3):289–296.
19. Schlueter A, Thesseling F. Building information model based energy/exergy performance assessment in early design stages. *Automation in Construction*. 2009;18(2):153–163.
20. Isikdag U, Underwood J. Two design patterns for facilitating Building Information Model-based synchronous collaboration. *Automation in Construction*. 2010;19(5):544–553.
21. Shen W, Shen Q, Sun Q. Building Information Modeling-based user activity simulation and evaluation method for improving designer-user communications. *Automation in Construction*. 2012;21(1):148–160.
22. Lee G, Park HK, Won J. D3 City project—economic impact of BIM-assisted design validation.*Automation in Construction*. 2012;22:577–586.

23. Lee S-I, Bae J-S, Cho YS. Efficiency analysis of Set-based Design with structural building information modeling (S-BIM) on high-rise building structures. *Automation in Construction*. 2012;23:20–32.
24. Lee J-K, Lee J, Jeong Y-S, et al. Development of space database for automated building design review systems. *Automation in Construction*. 2012;24:203–212.
25. Whyte J, Bouchlaghem N, Thorpe A, McCaffer R. A survey of CAD and virtual reality within the house building industry. *Engineering, Construction and Architectural Management*. 1999;6(4):371–379.
26. Mah D, Manrique JD, Yu H, Al-Hussein M, Nasseri R. House construction CODN2/DN footprint quantification: a BIM approach. *Construction Innovation*. 2011;11(2):161–178.
27. Sanguinetti P, Abdelmohsen S, Lee J, Lee J, Sheward H, Eastman C. General system architecture for BIM: an integrated approach for design and analysis. *Advanced Engineering Informatics*. 2012;26(2):317–333.
28. Hardin B. *BIM and Construction Management: Proven Tools, Methods, and Workflows*. New Jersey, NJ, USA: Sybex; 2009.
29. Davies R, Harty C. Implementing "Site BIM": a case study of ICT innovation on a large hospital project.*Automation in Construction*. 2013;30:15–24.
30. Elbeltagi E, Dawood M. Integrated visualized time control system for repetitive construction projects.*Automation in Construction*. 2011;20(7):940–953.
31. Azhar S. Building information modeling (BIM): trends, benefits, risks, and challenges for the AEC industry. *Leadership and Management in Engineering*. 2011;11(3):241–252.
32. Ding LY, Zhou Y, Luo HB, Wu XG. Using nD technology to develop an integrated construction management system for city rail transit construction. *Automation in Construction*. 2012;21(1):64–73.
33. Hartmann T, Van Meerveld H, Vossebeld N, Adriaanse A. Aligning building information model tools and construction management methods. *Automation in Construction*. 2012;22:605–613.
34. Ibrahim YM, Lukins TC, Zhang X, Trucco E, Kaka AP. Towards automated progress assessment of workpackage components

in construction projects using computer vision. *Advanced Engineering Informatics*. 2009;23(1):93–103.

35. Ren Y, Skibniewski MJ, Jiang S. Building information modeling integrated with electronic commerce material procurement and supplier performance management system. *Journal of Civil Engineering and Management*. 2012;18(5):642–654.

36. CRC for Construction Innovation. Adopting BIM for facilities management: solutions for managing the Sydney opera house. 2007, http://eprints.qut.edu.au/27582/1/27582.pdf.

37. Klein L, Li N, Becerik-Gerber B. Imaged-based verification of as-built documentation of operational buildings. *Automation in Construction*. 2012;21(1):161–171.

38. Becerik-Gerber B, Jazizadeh F, Li N, Calis G. Application areas and data requirements for BIM-enabled facilities management. *Journal of Construction Engineering and Management*. 2012;138(3):431–442.

39. Schevers H, Mitchell J, Akhurst P, et al. Towards digital facility modelling for Sydney Opera House using IFC and semantic web technology. *Electronic Journal of Information Technology in Construction*.2007;12:347–362.

40. Motamedi A, Hammad A. Lifecycle management of facilities components using radio frequency identification and building information model. *Electronic Journal of Information Technology in Construction*. 2009;14:238–262.

41. Wang Y, Wang X, Wang J, Yung P, Jun G. Engagement of facilities management in design stage through BIM: framework and a case study. *Advances in Civil Engineering*. 2013;2013:8 pages.189105

42. Jiao Y, Wang Y, Zhang S, Li Y, Yang B, Yuan L. A cloud approach to unified lifecycle data management in architecture, engineering, construction and facilities management: integrating BIMs and SNS.*Advanced Engineering Informatics*. 2013; 27(2):173–188.

Chapter 7

# Enhancing Knowledge Sharing Management Using Bim Technology in Construction

Shih-Ping Ho, Hui-Ping Tserng, and Shu-Hui Jan

Department of Civil Engineering, National Taiwan University, Number 1, Setion 4, Taipei 10617, Taiwan

## ABSTRACT

Construction knowledge can be communicated and reused among project managers and jobsite engineers to alleviate problems on a construction jobsite and reduce the time and cost of solving problems related to constructability. This paper proposes a new methodology for the sharing of construction knowledge by using Building Information Modeling (BIM) technology. The main characteristics of BIM include illustrating 3D CAD-based presentations and keeping information in a digital format and facilitation of easy updating and transfer of information in the BIM environment. Using the BIM technology, project managers and engineers can gain knowledge related to BIM and obtain feedback provided by jobsite engineers for future reference. This study addresses the application of knowledge sharing management using BIM technology and proposes a BIM-based Knowledge Sharing Management (BIMKSM) system for project managers and engineers. The BIMKSM system is then applied in a selected case study of a construction project in Taiwan to demonstrate the effectiveness of sharing knowledge in the BIM environment. The results demonstrate that the BIMKSM system can be used as a visual BIM-based knowledge sharing management platform by utilizing the BIM technology.

# INTRODUCTION

It is vitally important for project managers and jobsite engineers to obtain knowledge about construction and to solve any problems that may arise. To achieve this knowledge, jobsite engineers can learn from the experience of other jobsite engineers. Construction experience transfer involves using knowledge gained during the completion of previous projects to maximize the achievement of current project objectives [1]. In order to share knowledge between similar projects, construction professionals have traditionally used techniques ranging from annual meetings to face-to-face interviews [1]. In addition to experts' memory, construction experience can be recorded in various media, such as documents, databases, and intranets. Knowledge management (KM) is the collection of processes controlling the creation, storage, reuse, evaluation, and use of experience-based knowledge in a particular situation or problem-solving context. In construction, KM focuses on the acquisition and management of important experience-based knowledge provided by job engineers.

Regardless of whether a project executed by an architectural firm is successful, valuable knowledge can be gained and should be documented so that jobsite engineers can identify what worked and what did not. From the perspective of KM in construction, these experiences and the knowledge gained from them are valuable, as they are accumulated through large investments in manpower, time, and money. Most jobsite engineers agree that KM in construction projects is a vital tool construction management. The sharing of knowledge and feedback provided by jobsite engineers help to prevent mistakes that have been made in previous projects. Drawing on knowledge and experience thus eliminates the need to solve many problems from scratch.

Most recent construction projects in Taiwan have applied KM systems to improve construction management during the construction phase. However, most of the shared information during the construction phase is in the form of text-based information, with less focus on virtual illustration and sharing. In construction projects, KM may involve many important relationships between the presentation and retrieval of knowledge and CAD. Furthermore, when knowledge is available for sharing, it is not easy for engineers to understand it directly without 2D or 3D CAD illustrations. Building Information Modeling (BIM) is the process of generating and managing data during the building life cycle [2]. BIM technology has the potential to enable fundamental changes in project delivery that will support a more integrated, efficient process [3]. BIM digitally contains precise geometry and relevant data needed to support the design, procurement, fabrication, and construction activities used to build 3D object-oriented CAD [4]. The main characteristics of BIM include illustrating 3D CAD-based presentations, keeping information in digital format, and facilitating the easy updating and transfer of information in a BIM environment.

The primary purpose of the study is to develop a way for jobsite engineers to effectively acquire, manage, and reuse knowledge gained from other jobsite engineers and integrated using the BIM technology within the 3D CAD environment. The study proposes a novel approach using BIM-based knowledge models integrated with the BIM technology to track and manage valuable experience-based knowledge. The main function of the BIM technology in this study is the 3D BIM-based illustration of experience-based knowledge. The proposed BIM-based animated illustration of knowledge is applied to keep and explain information in a digital format and to facilitate the updating and transfer of knowledge in the BIM environment. Using BIM-based animated illustrations of information, engineers can get an overview of previous scenarios in selected projects and use the knowledge gained for future construction work and understand the setup/process from their own professional perspective. Project managers and jobsite engineers can track and access the most recently shared information on relevant issues during the construction phase. Knowledge of various issues can be updated quickly and made available to each participant in the visual environment. This study addresses the application of knowledge sharing management and proposes a BIM-based Knowledge Sharing Management (BIMKSM) system for project managers and engineers. The research is a pilot study that involves applying the BIMKSM system for knowledge management to a building project in Taiwan (see Figure 1), in order to analyze and discuss the whole process of construction knowledge sharing management and to implement the KM sharing during the construction phase. The processing and content of construction sharing experience-based knowledge can be modified according to a project's particular characteristics. Finally, the BIMKSM system is then applied in a selected case study to verify the effectiveness of sharing knowledge in the BIM environment.

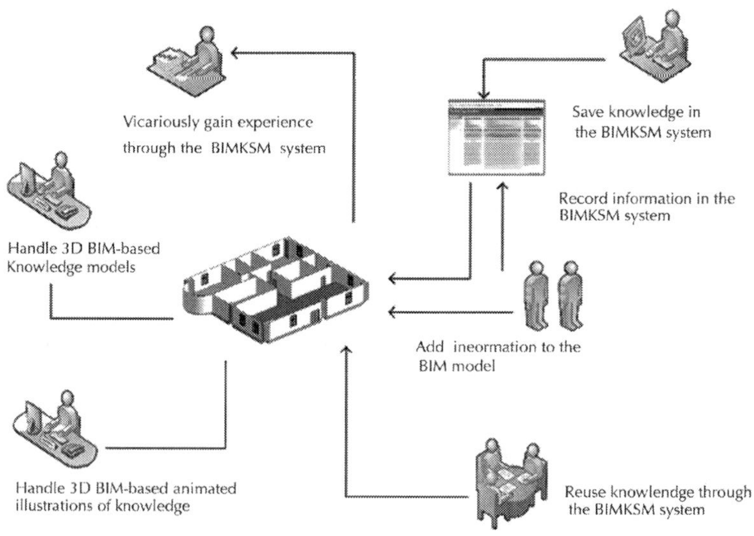


**Figure 1**: The application of the visual knowledge management in construction.

In sum, the purpose of this study is to (1) help jobsite engineers collect and share knowledge effectively through the BIMKSM system during the construction phase; (2) help engineers to refer to and exchange knowledge from the jobsite engineers using the BIMKSM system during the construction phase; (3) develop a web-based KM system to improve the exchange and tracking of knowledge for project managers and jobsite engineers.

# BACKGROUND RESEARCH

## Knowledge Management in Construction

Knowledge management deals with collecting, modeling, storing, reusing, evaluating, and maintaining knowledge [5–7]. Numerous research efforts have focused on applications of knowledge management in construction. El-Diraby and Kashif [8] presented a distributed ontology architectural design developed by rigorous knowledge acquisition and ontology development techniques for KM in the highway construction industry. Hartmann and Fischer [9] described how project teams can use 3D/4D models efficiently to support the communication of knowledge during the constructability review on construction projects. Ribeiro [10] analyzed KM effort based on case studies and provided recommendations and insights for enhancing KM in construction firms. Chen and Mohamed [11] provided empirical evidence for the stronger strategic role of tacit KM in comparison to explicit KM. Kivrak et al. [12] used a survey to find out how tacit and explicit knowledge are captured, stored, shared, and used in forthcoming projects and to identify major drivers and barriers in knowledge management. Chen et al. [13] presented a knowledge-sharing model to determine whether risk mitigation based on the use of derivatives would be beneficial to the companies. Forcada et al. [14] presented a survey of perceptions of KM implementation in the Spanish construction sector and compares the results obtained from design and construction firms.

## BIM Application in Construction

A great deal of previous research pertains to BIM issues in construction. Tse et al. [15] discussed the core barriers and recommended using BIM technology for construction industries. Mah et al. [16] proposed rapid computations of CO2 emissions from various house sizes, designs, and materials integrated with BIM technology. Goedert and Meadati [17] extended BIM technology into the construction process to create a single repository of facility data for the owner. Succar [18] explored publicly available international guidelines and introduced the BIM framework as a research and delivery foundation for industry stakeholders. Dossick and Neff [19] examined the use of BIM technologies for mechanical, electrical, plumbing, and life safety systems. Ren et al. [20] proposed a framework for integrating BIM for quantity takeoff and cost-estimating applications with e-commerce solutions for material procurement and supplier performance evaluation. Davies and Harty [21] developed BIM-enabled tools to allow site workers using mobile tablet personal computers to access design information and to capture work quality and progress data on-site. Martins and Monteiro [22] developed BIM-based automated code-checking procedures and system for water distribution systems. Zhang et al. [23] developed automated safety checking platform integrated BIM that informs construction engineers and managers for preventing fall-related accidents before construction starts. Wong et al. [24] highlighted critical initiatives derived from the review of BIM implementations in both the public and private sectors in six selected countries. Bryde et al. [25] explored the extent to which the use of BIM has resulted in reported benefits on a cross-section of construction projects. Succar et al. [26] proposed the five BIM framework components for the design, construction, and operation (DCO) stakeholders to measure and improve their BIM performance. Sebastian and Van Berlo [27] generated an instrument for benchmarking BIM performance to provide insight into the current BIM performance level of design, engineering, and construction firms. Bynum et al. [28] investigated the perceptions of the use of BIM for sustainable design and construction among designers and constructors. Wang et al. [29] explored how BIM will beneficially support facility management in the design phase, such as space planning and energy analysis. Zhang et al. [30] proposed and verified Industrial Foundation Classes-based graphic information model as the foundation of data sharing in virtual construction systems and in other AEC/FM applications.

Although numerous knowledge management systems have been developed for the application of construction knowledge management, such systems typically exist for knowledge sharing using only text-based illustrations. To enhance construction-related knowledge sharing using a BIM-based environment, this study proposes a novel management system for project managers and jobsite engineers.

# METHODOLOGY

## Bim-Based Knowledge Management Approach

In this study, the proposed BIMKSM system facilitates visual knowledge sharing and management using the BIM technology during the construction phase. The BIM technology stores any problems, solutions, and comments, allowing project managers and jobsite engineers to access the most up-to-date knowledge. The primary advantages of the proposed BIMKSM system are (1) to effectively link knowledge using BIM-based graphic representations; (2) to promote relationships between areas of expertise via both vertical and horizontal graphic representations; and (3) to provide statuses of acquired knowledge of different situations using different colors.

Most knowledge can be classified as either tacit or explicit knowledge. Tacit knowledge is personal, context-specific knowledge that is difficult to formalize, record, or articulate. This type of knowledge is stored in the heads of people [31]. Tacit knowledge or experience is primarily developed through a process of trial and error in practice. Tacit knowledge that can be communicated directly and effectively is personal knowledge embedded in individual experience and shared and exchanged through direct, face-to-face contact [32]. In contrast, the acquisition of explicit knowledge is indirect; it must be decoded and recoded into one's mental models and is then internalized as tacit knowledge. Explicit knowledge can be codified and transmitted in a systematic and formal language. Explicit knowledge can be found in the documents of organizations, including reports, articles, manuals, patents, pictures, images, video, audio, software, and other forms. In this study, "tacit knowledge" refers to "hard" information that is visibly or invisibly related to part of an area of knowledge, including experience and know-how. Explicit knowledge is "soft" information that enables or facilitates the execution of specific information, including contracting, drawing, solving problems, or approving proposals. All jobsite engineers are responsible for sharing knowledge pertaining to their own domain. Any BIM model whose integrated information/knowledge sharing requirements have been noted will be classified as explicit in order to allow relevant experiences and processes to be recorded. Therefore, the shared information associated with objects of BIM model can be referred to and reused in other projects.

Shared information from all jobsite engineers is divided and saved as "activity," "object," or "issue" for collection and management. The main advantage of BIM-based knowledge management is the ease with which information and knowledge can be understood and reapplied. Knowledge saved in the "issue" category includes both tacit and explicit knowledge. With respect to explicit knowledge, BIM-related information normally includes original comments, reports, drawings, documents, and comments submitted by jobsite engineers. In contrast, tacit knowledge may include process records, problems faced, problems solved, expert suggestions, know-how, innovations, and notes on experience. Such information is better saved in issue-based components to facilitate classification and searching by users. Information that relates to the whole project that cannot be easily classified into issue components is saved under the "project" category.

A BIM-based knowledge model can be defined as a graphic representation of experiences linking relationships between objects of the BIM model and aspects of experience-based knowledge. The BIM technology retains knowledge in a digital format, facilitating easy updating and transfer of knowledge into the BIM environment. A BIM-based knowledge model is designed to be easily integrated with experience-based information and objects of the model (see Figure 2). Information in the BIM-based knowledge models can be identified, tracked, and managed, and problems encountered during construction projects can be solved. The most up-to-date knowledge and solutions can be acquired from participating engineers and then shared and saved as objects of the BIM model for future reference. The model is constructed from variables that can be decomposed into objects of a BIM model and can then store the identified knowledge. Information stored in the objects of BIM models includes both facing problems and solutions. Facing problems may be knowledge issues, knowledge attributes, descriptions of problems, or knowledge attachments (e.g., documents, reports, drawings, and photographs).

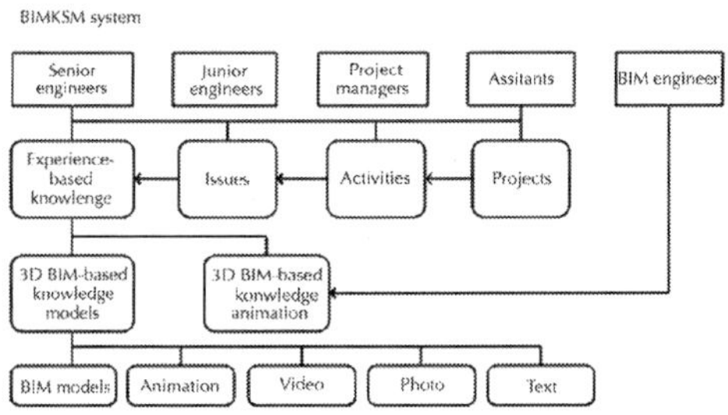


**Figure 2:** The concept and framework of the BIM-based knowledge models.

## Procedures of BIM-Based Knowledge Models Usage

The procedures for using BIM-based knowledge models are based on a knowledge management framework. The procedure consists of three phases: creating an issue, sharing knowledge, and updating said knowledge. Figure 3 presents the flowchart of the procedure of the knowledge models usage for knowledge sharing.

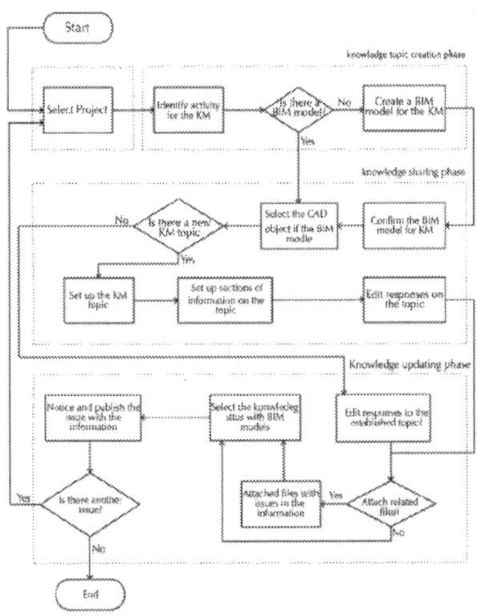


**Figure 3**: A flowchart of the procedure of the knowledge models usage.

## ssue Creation Phase

The initial engineer may determine which projects, activities, and issues are suitable for knowledge sharing. Furthermore, the issue must be set up by the initial participant (engineer) at the beginning of the phase. Such information under knowledge issues includes determining the type of knowledge, objects of BIM model, activities, and projects that should be assigned in association with the issue.

## Knowledge Sharing Phase

After studying the published materials, all qualified and interested engineers are invited to edit and submit any knowledgeable comments they may have on the issue. All explicit knowledge prepared by engineers needs to be digitized by them or by assistants before it can be submitted to the BIMKSM system. All knowledge must also be examined and confirmed before publishing. All interested engineers can discuss problems related to selected issues and objects of the BIM model and seek responses from other engineers and managers through the BIMKSM system. Meanwhile, the engineers can direct responses either to individuals or a group. After tacit and explicit knowledge is saved in the system, all knowledge can be referenced and reused. Engineers can gain knowledge from the issues catalogue of the objects and can access this catalogue for use in other similar projects.

## Knowledge Updating Phase

After applying tacit and explicit knowledge to similar projects, the engineers can resolve their problems related to those issues. Finally, the engineers can note and submit the new tacit knowledge and record the experiences through which it was gained, and they can associate this information with the original knowledge. Furthermore, the information is updated again because further feedback and updated knowledge are provided regarding the issues. After the approval process has been completed, the updated knowledge set is republished to authorized members.

# SYSTEM IMPLEMENTATION

## System Architecture

The BIMKSM system provides a user friendly portal for all project participants. It also serves as a real-time online communication channel for knowledge management. All data are stored and classified using BIM-based knowledge models. The BIMKSM system is a solution that uses a single unified database linked to BIM files with different levels of access granted to users based on their roles. Authorized participants can access the BIM-based knowledge models to update information on content relevant to the user's responsibilities in the project. When information is updated in the BIMKSM system, the server automatically sends e-mails to the project managers and jobsite engineers involved in the project.

The developed BIMKSM system runs on Microsoft Windows 2008 software with an Internet Information Server (IIS) as the web server. The BIMKSM system was developed using Java Server Pages (JSPs), which are easily incorporated with HTML and JavaScript technologies. The BIMKSM system server supports four distinct layers: interface, access, application, and database layers. Each layer has its own responsibilities. The interface layer defines administrative and end-user interfaces. Users can access information via web browsers such as Microsoft Internet Explorer or Google Chrome. Administrators control and manage information via the web browser or a separate server interface. The access layer provides system security, restricted access, firewall services, and system administration functions. The application layer defines various applications for analyzing and managing information. The database layer consists of a primary Microsoft SQL Server 2003 database. A firewall and virus scan can be used to protect the system database against intrusion. Users can utilize the BIM models in the BIMKSM system to request assistance or send word of a problem directly to the BIMKSM system to ask for further support.

In this study, the BIM model is interpreted as an information model in the BIMKSM system. The application of BIM models to acquire and store information about an object involve a description of the problem being faced, knowledge, comments, and attaching documents in the

BIMKSM system. Autodesk Revit Architecture and Revit MEP were used to create the BIM model and files. Autodesk Design Review was used to read the files for BIM-based knowledge models. Also, Autodesk Inventor was used to create BIM-based animations to illustrate the knowledge. The integration of the information with the BIM models was achieved using the Autodesk Revit application programming interface (API) and the Microsoft Visual Basic. Net (VB.Net) programming language. The BIM-based knowledge models were developed in Autodesk Revit Architecture and Revit MEP by programming in VB.Net and using Revit API. All information in the BIM files could be exported to an ODBC database for connection with the BIMKSM system. Figure 4 shows the BIM-based knowledge sharing and management process flowchart in the BIMKSM system.

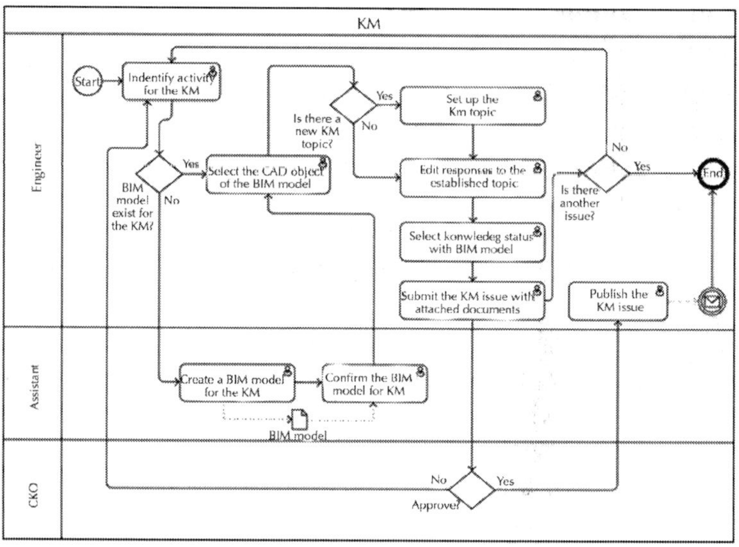


**Figure 4**: The BIM-based knowledge sharing and management process flow-chart.

## System Modules

All modules in the system are briefly described below.

## Authority Setup Module

The authority setup module is an access control mechanism preventing unauthorized users from entering the system or retrieving sensitive information. The BIMKSM system requires all project participants (managers and jobsite engineers) to register by providing a unique user ID and password for authentication.

## Knowledge Edition Module

Through this module, project participants can edit their relevant contributions to the objects in the BIM model. Generally, participants may create a new issue or contribute to those started by others in order to share their knowledge on various aspects of the project. The edited information will be saved in issues by categories associated with the relevant objects of the BIM model. Also, attached documents and report files must be uploaded in PDF format, the standard file format. The knowledge edition module allows experts and engineers to share issue-based tacit knowledge via a discussion forum.

## Alert Setup Module

This module helps project participants set up an alert service for monitoring knowledge via e-mail. Dates related to any notification of new information are submitted or updated systematically and project participants can determine who is invited to submit knowledge.

## Report Module

The report module allows users to easily access the summarized information to identify needs and analyze what has been recorded. The knowledge report can be illustrated with a BIM model, a description, and a summary of the information. Furthermore, all reports can be presented on the web or extracted to PDF format. This process allows users to make and organize knowledge-related reports from a central location.

# CASE STUDY

## Case Study Introduction

In the following case, the architecture firm has had sixteen years of experience specifically in construction building projects. The architecture firm hoped to take full advantage of the BIM-based knowledge management system to facilitate knowledge exchange and management during the construction phase and reuse it in other similar projects. Therefore, the architecture firm announced that all jobsite and project managers would be encouraged to use the BIMKSM system to apply knowledge management in order to effectively manage acquired information during the construction phase in the BIM environment. The BIMKSM system was utilized in the construction project to verify the proposed methodology and demonstrate the effectiveness of sharing previous experience in the construction phase. The project used as a case study lasted 4 months. During the study, all jobsite and project managers were encouraged to explore and edit their own recorded experiences in the BIMKSM system. Figure 5 shows the BIM-based knowledge animation usage process. Figure 6 illustrates application of BIM-based knowledge models for knowledge sharing in the case study.

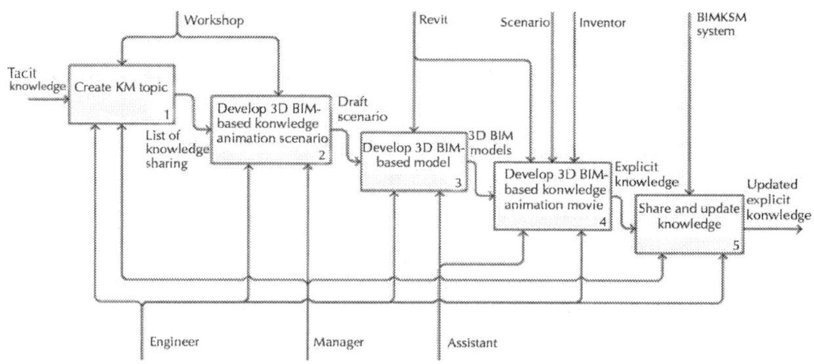


**Figure 5:** The BIM-based knowledge animation usage process.

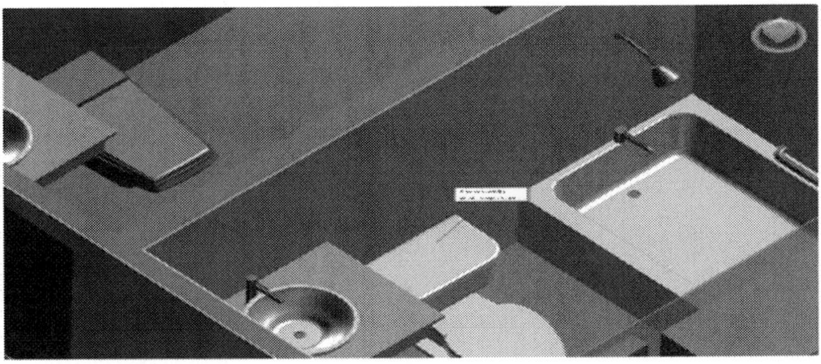

(a)

(b)

**Figure 6**: Application of BIM-based knowledge models for knowledge sharing in the case study.

As previously mentioned, the case study was implemented in the middle of the construction phase. All BIM models used in the study were created and developed for the purpose of construction management. Finally, the BIM models were reused and applied for knowledge management.

First, the engineers were invited to explain their experiences and provide comments based on the issue and include relevant information and documents. The initial engineer created issues regarding the selected activity and objects of the BIM model in the initial phase. All related documents for this issue were collected and digitized by the senior engineers and knowledge assistants. After the issue was created, the senior engineers were invited to share their knowledge and comments related to the issue using the system. The posted files included digital documents, photos, and films. The knowledge assistants helped the senior engineers to digitize the content, and they then created the BIM objects related to the issue. The other issues were communicated in the same manner. All engineers were required to submit experience-based information and discussions regarding the issue via the BIMKSM system. The engineers read previous comments provided by others, learned from these records, and submitted their own comments via the BIMKSM system, which then allowed other engineers to discuss their work. The comments provided by the senior engineers included notes, actual problems/solutions, and suggestions. The engineers communicated problems and solutions to the senior engineers, posted their comments in the system, and shared their case discussions with others. The engineers were required to submit their knowledge pertaining to the BIM objects of the BIM model via the BIMKSM system. The senior engineers reviewed all questions and solutions and posted comments for all interested individuals. Furthermore, all information was stored in the central database to prevent the collection of redundant data. Finally, the information was automatically backed up from the system database to another database. Moreover, the knowledge was updated later because further feedback and another solution to the same problem were provided. The updated knowledge set was republished in the object of BIM model of the BIMKSM system after the approval process was completed. A notice message was then transmitted to authorized members.

## Case Study Evaluation

Questionnaire results from the case study evaluation reveal that the BIMKSM system effectively shares knowledge. A verification test was performed by checking whether the BIMKSM system could perform tasks as specified in the system analysis and design. A validation test was undertaken by requesting selected case project participants to use the system and then providing feedback by answering a questionnaire. There were 18 respondents: two project managers with ten years of experience; six jobsite engineers with ten years of experience; seven jobsite engineers with five years of experience; two assistants with three years of experience; and one CKO with ten years of experience.

The BIMKSM system was demonstrated to the respondents, who were requested to give their opinions on it by completing the questionnaire. Table 1 shows the results of the system evaluation.

**Table 1:** System evaluation results

| System evaluation item | Mean score |
|---|---|
| Enhances visual knowledge illustration | 4.5 |
| Applicability to construction knowledge management | 4.8 |
| Reduces rework percentage | 4.2 |
| Reduces percentage of mistakes occurring | 4.3 |
| Improves the knowledge collection | 4.2 |
| Enhances knowledge communication | 4.4 |
| Improves knowledge sharing | 4.7 |
| Enhances learning performance | 4.3 |

**Table note**

The mean score is calculated from respondents' feedback on a five-scale questionnaire: 1 (strongly disagree), 2, 3, 4, and 5 (strongly agree).

Overall, the jobsite engineers' feedback for the case study was positive. Most engineers and project managers agreed that the BIMKSM system helps them to view all collected knowledge and experience in the BIM environment. Questionnaire results indicate that the primary advantages of using the BIMKSM system are as follows: (1) it provides 3D visual illustration of knowledge regarding project-based knowledge (86% agreed); (2) it provides BIM-assisted animation easily and effectively (89% agreed); and (3) it clearly identifies available knowledge and different status of different knowledge in the BIM environment (90% agreed).

Questionnaire results indicate that the primary advantages of the application of BIM in BIM-based knowledge models are as follows: (1)

it provides clear 3D representations, thus identifying knowledge and experience feedback relevant to an object or activity (92% agreed); (2) one can generate a visual object of BIM model illustrations of knowledge, thus identifying acquired knowledge and experience feedback relevant to tasks and projects (90% agreed); (3) it allows one to view knowledge and information provided by jobsite engineers easily and effectively (90% agreed); and (4) it enables engineers to trace and manage any acquired BIM-based knowledge feedback (88% agreed).

User feedback indicated that the primary barriers to using the BIMKSM system were as follows: (1) insufficient updated information is related to different types of knowledge; (2) substantial amounts of time and assistance are needed for engineers and managers to use BIM software to edit and update knowledge feedback; (3) further effort is required to update information related to various objects of a BIM model or the activities in a project; (4) the senior and jobsite engineers require substantial time and assistance to edit knowledge feedback in BIM environment; and (5) few engineers do not accept BIM applications in the construction sites.

# CONCLUSION

To improve construction knowledge sharing in building projects, this work presented and developed the BIMKSM system as a visual platform. The BIMKSM system illustrates knowledge with problem descriptions and solutions in the BIM environment. BIM is a highly promising means of enhancing knowledge management during the construction phase of a project. Collecting problem descriptions and solutions using the BIM technology allows project managers and jobsite engineers to contribute and share the most up-to-date knowledge and experience regarding problems and solutions in construction. The BIM technology generates 3D drawings, thus identifying valued experience-based knowledge relevant to issues and activities. Additionally, BIM provides objects and illustrations when knowledge is available. The BIMKSM system collects specific problem solutions and supports all information across projects. Overall, field test results indicate that the BIMKSM system is an effective and simple platform for knowledge management in construction projects. The case study results demonstrate the effectiveness of a BIMKSM-like system for KM due to its incorporation of BIM and web technologies during the construction phase.

The concept of a BIM-based knowledge management system was presented, as was a system for use as a knowledge-sharing platform in construction design-build projects. The application of BIM-based knowledge management system mainly allows project participants to access the knowledge easily and effectively. Although effort is required to update information on various problems and solutions, the proposed system benefits KM by (1) developing web-based BIM-assisted knowledge management system for construction knowledge sharing, (2) providing an effective and efficient method to assist and manage visual KM work, (3) enabling users to learn knowledge through BIM-based knowledge animation.

In sum, the engineers were able to increase their understanding of previous captured knowledge and experience from all participants working on different projects. Notably, BIM integrates the objects comprising knowledge management work by incorporating external factors, such as problem and solution descriptions, comments, and suggestions, into a single source for all construction BIM-based KM information. Effectively utilizing web technologies and BIM during the construction phase allows project participants to enhance knowledge sharing for domain knowledge management. The case studies also show that a BIMKSM system can provide a web-based visual BIM-assisted knowledge sharing management platform.

# ACKNOWLEDGMENTS

The authors would like to express their appreciation to S-D Construction Inc. for assistance in the system design and interviews and to the experts and engineers of the owner of the project for providing useful data, valuable information, and helpful comments during system design and development.

# CONFLICT OF INTERESTS

The authors declare that there is no conflict of interests regarding the publication of this paper.

# REFERENCES

1. Reuss MC, Tatum CB. No. 78. Stanford, Calif, USA: Center for Integrated Facility Engineering, Stanford University; 1993. Requirements and tools for transferring construction experience between projects.
2. Epstein E. Implementing Successful Building Information Modeling. Norwood, Mass, USA: Artech House; 2012.
3. Teicholz P. BIM for Facility Managers. New York, NY, USA: John Wiley & Sons; 2013.
4. Eastman C, Teicholz P, Sacks R, Liston K. BIM Handbook: a guide to Building Information Modeling for Owners, Managers, Designers, Engineers and Contractors. New York, NY, USA: John Wiley & Sons; 2008.
5. Davenport TH, Prusak L. Working Knowledge. New York, NY, USA: Harvard Business School Press; 1998.
6. Bergmann R. Knowledge Management: Foundations, Development Methodology, and Internet-Based Applications. Berlin, Germany: Springer; 2002.
7. Davenport TH, Probst GJB. Knowledge Management Case Book: Siemens Best Practices. 2nd edition. New York, NY, USA: John Wiley & Sons; 2002.
8. El-Diraby TE, Kashif KF. Distributed ontology architecture for knowledge management in highway construction. Journal of Construction Engineering and Management. 2005;131(5):591–603.
9. Hartmann T, Fischer M. Supporting the constructability review with 3D/4D models. Building Research and Information. 2007;35(1):70–80.
10. Ribeiro FL. Enhancing knowledge management in construction firms. Construction Innovation. 2009;9(3):268–284.
11. Chen L, Mohamed S. The strategic importance of tacit knowledge management activities in construction. Construction Innovation. 2010;10(2):138–163.
12. Kivrak S, Arslan G, Dikmen I, Birgonul MT. Capturing knowledge in construction projects: knowledge platform for contractors. Journal of Management in Engineering. 2008;24(2):87–95.

13. Chen JH, Hsu SC, Luo YH, Skibniewski MJ. Knowledge management for risk hedgingby construction material suppliers. Journal of Management in Engineering. 2012;28(3):8–26.
14. Forcada N, Fuertes A, Gangolells M, Casals M, Macarulla M. Knowledge management perceptions in construction and design companies. Automation in Construction. 2013;29:83–91.
15. Tse T-CK, Wong K-DA, Wong K-WF. The utilisation of building information models in nD modelling: a study of data interfacing and adoption barriers. Electronic Journal of Information Technology in Construction. 2005;10:85–110.
16. Mah D, Manrique JD, Yu H, Al-Hussein M, Nasseri R. House construction CO2 footprint quantification: a BIM approach. Construction Innovation. 2011;11(2):161–178.
17. Goedert JD, Meadati P. Integrating construction process documentation into building information modeling. Journal of Construction Engineering and Management. 2008;134(7):509–516.
18. Succar B. Building information modelling framework: a research and delivery foundation for industry stakeholders. Automation in Construction. 2009;18(3):357–375.
19. Dossick CS, Neff G. Organizational divisions in bim-enabled commercial construction. Journal of Construction Engineering and Management. 2010;136(4):459–467.
20. Ren Y, Skibniewski MJ, Jiang S. Building information modeling integrated with electronic commerce material procurement and supplier performance management system. Journal of Civil Engineering and Management. 2012;18(5):642–654.
21. Davies R, Harty C. Implementing 'Site BIM': a case study of ICT innovation on a large hospital project. Automation in Construction. 2013;30:15–24.
22. Martins JP, Monteiro A. LicA: a BIM based automated code-checking application for water distribution systems. Automation in Construction. 2013;29:12–23.
23. Zhang S, Teizer J, Lee JK, Eastman CM, Venugopal M. Building Information Modeling (BIM) and Safety: automatic safety checking of construction models and schedules. Automation in Construction. 2013;29:183–195.

24. Wong AKD, Wong FKW, Nadeem A. Attributes of building information modelling implementations in various countries. Architectural Engineering and Design Management. 2010;6:288–302.
25. Bryde D, Broquetas M, Volm JM. The project benefits of Building Information Modelling (BIM) International Journal of Project Management. 2013;31(7):971–980.
26. Succar B, Sher W, Williams A. Measuring BIM performance: five metrics. Architectural Engineering and Design Management. 2012;8(2):120–142.
27. Sebastian R, Van Berlo L. Tool for benchmarking BIM performance of design, engineering and construction firms in the Netherlands. Architectural Engineering and Design Management. 2010;6:254–263.
28. Bynum P, Issa R, Olbina S. Building information modeling in support of sustainable design and construction. Journal of Construction Engineering and Management. 2012;139(1):24–34.
29. Wang Y, Wang X, Wang J, Yung P, Jun G. Engagement of facilities management in design stage through BIM: framework and a case study. Advances in Civil Engineering. 2013;2013:8 pages.189105
30. Zhang J, Yu F, Li D, Hu Z. Development and implementation of an industry foundation classes-based graphic information model for virtual construction. Computer-Aided Civil and Infrastructure Engineering. 2012
31. Hart A. Knowledge Acquisition for Expert Systems. 2nd edition. New York, NY, USA: McGraw-Hill; 1992.
32. Tiwana A. The Knowledge Management Toolkit—Practical Techniques for Building a Knowledge management System. Upper Saddle River, NJ, USA: Prentice-Hall; 2000.

Chapter 8

# Developing Mobile Bim/2D Barcode-Based Automated Facility Management System

Yu-Cheng Lin, Yu-Chih Su, and Yen-Pei Chen

Department of Civil Engineering, National Taipei University of Technology, No. 1 Chung-Hsiao E. Road, Section 3, Taipei 10608, Taiwan.

## ABSTRACT

Facility management (FM) has become an important topic in research on the operation and maintenance phase. Managing the work of FM effectively is extremely difficult owing to the variety of environments. One of the difficulties is the performance of two-dimensional (2D) graphics when depicting facilities. Building information modeling

(BIM) uses precise geometry and relevant data to support the facilities depicted in three-dimensional (3D) object-oriented computer-aided design (CAD). This paper proposes a new and practical methodology with application to FM that uses an integrated 2D barcode and the BIM approach. Using 2D barcode and BIM technologies, this study proposes a mobile automated BIM-based facility management (BIMFM) system for FM staff in the operation and maintenance phase. The mobile automated BIMFM system is then applied in a selected case study of a commercial building project in Taiwan to verify the proposed methodology and demonstrate its effectiveness in FM practice. The combined results demonstrate that a BIMFM-like system can be an effective mobile automated FM tool. The advantage of the mobile automated BIMFM system lies not only in improving FM work efficiency for the FM staff but also in facilitating FM updates and transfers in the BIM environment.

# INTRODUCTION

Facility management (FM) during the operation and maintenance phase of a facility's lifecycle has become an important topic in research and academic studies. Managing the inspection and maintenance information of equipment and facilities contributes to successful FM. Managing the work of FM effectively during the operation and maintenance phase can be extremely difficult owing to the various types of equipment and facilities. Furthermore, it is inconvenient for FM staff to maintain those facilities by relying on paper-based documents. Unlike the manufacturing industry, information technology is limited in its use and its application in construction [1], with human labor conducting most of the management work, which is inefficient and sometimes error-prone [2].

Building information modeling (BIM) is a computable representation of all of a building's physical and functional characteristics and related lifecycle information and serves as a repository of information for building owners and operators that is used and maintained throughout the lifecycle of a building [3]. BIM is an emerging visual communication tool in the architecture, engineering, and construction (AEC) industry. Recently, various BIM applications have been applied during design and construction phases. However, without a maintenance stage application, BIM cannot fulfill the "lifecycle" mission. Although many projects have been implemented for FM with the use of BIM technology, problems and challenges remain in applied BIM technology that need to be solved and improved in practice.

In FM, the staff usually refers to information such as specifications, checklists, maintenance reports, and maintenance records. As FM staff must record inspection and maintenance results in hard copies, there can consequently be significant gaps in data capture and entry. Such means of communicating information between the facility location and the management office are ineffective and inconvenient. According to the survey findings regarding maintenance work on a commercial building in Taiwan [4], the primary problems regarding data capture and sharing during the FM process are as follows: (1) the efficiency and quality are low, especially through document-based media; (2) it is not easy to refer to the relevant detailed information on facilities; (3) there are data reentry problems; and (4) the use of desktops for operating the BIM models cannot be effectively extended to maintenance management services at the facility location. However, few suitable platforms exist to assist FM staff in using an integrated FM information system from the BIM models and in sharing maintenance information directly at the facility's location.

The performance of FM can be enhanced by using Internet technology for information-sharing and communication. In this study, the work of FM includes inspection and maintenance work. By integrating automatic identification technologies (such as two-dimensional (2D) barcode systems), the effectiveness of FM work is enhanced and improved (see Figure 1). In order to enhance the effectiveness of FM work on commercial buildings, this study presents a novel system called the mobile automated BIM-based facility management (BIMFM) system for the acquisition and tracking of maintenance information and provides an information-sharing platform for FM staff that may be accessed with the use of a webcam-enabled notebook or tablet. Integrating BIM and 2D barcode technologies, information, and data entry mechanisms can help to improve the effectiveness and convenience of the information flow in the FM process. The primary objectives of this study include (1) applying BIM and 2D barcode technologies to increase the efficiency of FM data and information collection, (2) directly accessing 2D barcode technologies to link detailed information to the BIM models of facilities, (3) developing a mobile BIM/2D barcode-based system to assist directly the BIM-based maintenance management work at facility locations, and (4) exploring the limitations of the system, addressing problems, and providing suggestions based on the implementation of the case study. The mobile automated BIMFM system is applied to a commercial building in Taiwan to verify our proposed methodology and demonstrate the effectiveness of the FM process in construction. There are two hypotheses in this study: the first is that all BIM models are developed during the construction phase and made ready for FM; the second is that all BIM models must be updated and corrected constantly.

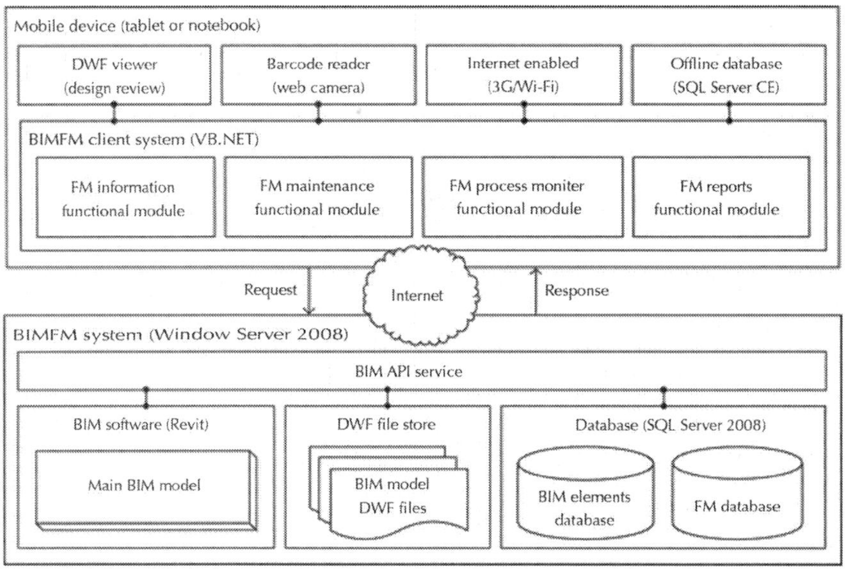


**Figure 1:** Overview of the BIMFM system framework.

## Related Research Studies

A substantial amount of research has shown the potential of one-dimensional barcode applications in various areas of the construction industry, such as data entry efficiency, labor management, productivity improvement, cost savings, construction equipment and materials tracking, and electronic document management [5–9]. McCullouch and Luepraser [10], for example, illustrated how 2D barcode technology could be applied in the construction industry. Various other research works on the application of barcode models have focused on the integration of other technologies. Navon and Berkovich [11], for example, used barcode and radio frequency identification (RFID) technologies for automated data collection to assist with materials management and control. Shehab and Moselhi [12] illustrated the use of barcode technology to develop an automated system for retrieving engineering deliverables such as drawings, reports, and specifications. Saeed et al. [13] integrated a global positioning system (GPS) with RFID and 2D barcode technologies to provide a solution for pedestrian users that allowed them to access information about buildings and other artifacts.

BIM is changing the traditional construction practices in a broader sense in terms of people, process, working culture, communication, business models, and so forth [14]. Many core benefits, barriers, frameworks, and recommendations for BIM usage are cited in previous work on supporting decisions and improving processes throughout the lifecycle of a project [3, 15–24]. A substantial amount of previous research has examined BIM issues in the operation phase of construction. The Sydney Opera House adopted BIM technology as a means of support for their integrated facility management [25]. Motamedi et al. [26] utilized BIM visualization capabilities to provide FM technicians with visualization that allowed them to utilize their cognitive and perceptual reasoning for problem solving. Becerik-Gerber et al. [27] assessed the status of BIM implementations in FM, the potential applications, and the level of interest in the utilization of BIM by conducting online surveys and face-to-face interviews. Wang et al. [28] not only developed a framework through which one could consider FM in the design stage through BIM but also explored how BIM would beneficially support FM in the design phase. Lin et al. [29] processed different kinds of building components and their corresponding properties to obtain rich semantic information that could enhance applications of path planning in FM. Costin et al. [30] utilized RFID technology for real-time visualization and location tracking in a BIM model. Gheisari et al. [31] explored the ways through which one could integrate BIM with mobile augmented reality (MAR) and make the data accessible through handheld mobile devices in order to enhance current facility management practices.

The BIM approach, which is used to retain facility information in a digital format, facilitates easy updates of FM information in a BIM environment. Although there are many practical applications for using BIM in the maintenance management stage, there are challenges as well. One of the challenges involves the accessibility of the BIM models for FM staff: it usually takes time to refer to and link the corresponding FM element in the BIM model during the maintenance and inspection process [32]. To assist FM staff in obtaining the corresponding BIM model for facilities maintenance management in an automatic and effective manner, this study develops a proposed system that integrates 2D barcode technology to connect automatically to the BIM models. This study then manages facilities by using the 2D barcode technology that is integrated with the BIM approach. By using 2D barcode technology, users can link to the corresponding BIM model of a facility in a quick, automatic, and effective way and access basic information and maintenance problems, while managing FM information during the operation and maintenance phase. Next, the proposed BIMFM system is applied to a case study of a commercial building project in Taiwan to verify its efficacy and demonstrate its FM effectiveness in a BIM-based environment. Finally, the limitations, problems, and suggestions are discussed based on the implementation of the case studies in this study.

# KEY TECHNOLOGIES

## BIM Technology

BIM is one of the most promising recent developments in the AEC industry [33]. It was developed nearly ten years ago with the aim of providing an environment from which any related information on three-dimensional (3D) entity models could be retrieved during the project lifecycle [14, 34]. BIM is considered essential in AEC for the management, sharing, and exchange of information among project stakeholders such as architects, engineers, contractors, owners, and subcontractors [35], although its technologies are being adopted more slowly in the AEC industry than 2D computer-aided design (CAD) [36, 37]. By enabling visualization of the details of the prospective work,

BIM assists construction planners in making crucial decisions [38]. BIM is a new technology in the field of CAD, which contains not only geometric data but also a great amount of engineering data throughout the lifecycle of a building [39]. As a digital tool, BIM supports the continual updating and sharing of project design information [30]. A BIM system enables users to integrate and reuse building information and domain knowledge throughout the lifecycle of a building [40].

There are many BIM commerce tools for creating BIM models (e.g., Autodesk Revit, Trimble Tekla, and the Graphisoft ArchiCAD software). Most of these commerce tools provide software development kits (SDK) for programming purposes. For this study, Autodesk Revit is selected as the main BIM tool because it provides more SDK support than other commerce tools. Furthermore, the use of Autodesk Revit allows the easy export of all of the information regarding a BIM model to a database through open database connectivity (ODBC).

## 2D Barcode Technology

Another technology explored as a means of providing accurate and reliable real-time inspection information is the 2D barcode system. The 2D barcode system also has the ability to deliver information on location, including text, audio, and video. Barcode technology was invented in 1950 and it developed rapidly during the subsequent years. With the advantages of higher capacity, lower cost, increased security, traceability, anticorruptibility, and mistake-correcting functionality, the 2D barcode has been widely applied since 1990 [9]. The major characteristics of the 2D barcode are its capacity to represent data content and its arrangement of a specific geometric diagram in a relatively small matrix area that can record significant quantities of data. The 2D Stacked Code and the 2D Matrix Code are the two typical types of barcode classified by their design principle. The 2D Stacked Code was developed based on the one-dimensional barcode. It is composed by thinning down the one-dimensional barcode and stacking it in layers to create multirow symbols. Representative types of the Stacked Code include Code 16K, Code 49, and Portable Data File 417 (PDF417). The 2D Matrix Code was composed by the distribution of black-white picture elements (square, dot, or other types) in a square area in relative matrix position. Representative types of Matrix Codes include code one, maxi code, quick response (QR) code, and data matrix.

Although many types of 2D barcodes exist, as shown above, the QR code is the most popular type of 2D barcode used in Taiwan. The advantages of the QR code are as follows:(1)high capacity of data content: the QR code can record thousands of characters or numbers, since its capacity is ten times greater than that of the one-dimensional barcode;(2)various data types: the data types stored in the 2D barcode include image, sound, words, and fingerprints, with the capacity for multilanguage expression;(3)ease of production: the scale and shape of the QR code are changeable and are easily made by software and a printer, at a low cost;(4)convenience: the QR code can easily be identified by a mobile phone or a mobile device and is readable in any direction (http://www.qrcode.com/en/about/).

Although RFID technology is suitable for long-distance reading in FM work, the cost of RFID readers and tags is a major problem when many readers and tags are needed. Furthermore, tablets have free software for reading the QR code. Therefore, the QR code is selected and utilized in this study because QR code labels are cheaper than RFID tags.

## SYSTEM SCHEMATIC DESIGN

The 2D barcode has been widely applied in Taiwan. With the advancement of mobile technology, many mobile phones are equipped with cameras, which have the capability to scan 2D barcodes. When a barcode reader program is installed, the user can quickly access product descriptions, web addresses, or e-mail addresses by scanning the barcode. For example, a mobile phone with the Google Android system can read one- or two-dimensional barcodes such as the international article number (EAN), the international standard book number (ISBN), or the QR code after installing the Zxing barcode scanner software. Tablets equipped with cameras also enable the application of 2D barcode scanning to facilitate maintenance management of building facilities. The 2D barcode can be easily identified by mobile devices and record thousands of characters or numbers, since its capacity is ten times greater than the one-dimensional barcode. Furthermore, the 2D barcode's ability to decode mistakes is much higher than that of the one-dimensional barcode [9]. In this study, the main reason for using the 2D barcode is that the brief information and the uniform resource locator (URL) for directly linking the BIM models can be stored within the 2D barcode, unlike the one-dimensional barcode. It is an easy and effective way to link to the BIM models. In this study, we do not adopt an RFID solution because the use of RFID requires RFID tags and an additional RFID reader hardware. The purchasing cost of the RFID tags and the additional RFID reader hardware would be higher than the cost of the 2D barcodes used in this study. Furthermore, 2D barcode labels can be printed without the use of a specialized printer and can be scanned and read by webcam-enabled tablets. Based on the considerations of cost and effectiveness, the 2D barcode is a better choice for implementation. Therefore, this study integrates BIM and 2D barcode technologies to enhance FM work and provide detailed FM information communication. An integrated client/server platform can link all of the information on building facilities to improve the effectiveness of the FM process.

The application of BIM/2D barcode technology in the management of facilities both inside and outside of the buildings focuses on its rapid identification and supports FM staff in handling FM via the 3D BIM models. By scanning the 2D barcode label sticker on a facility, FM staff can obtain the corresponding BIM model of the facility and directly access FM information about the facility, such as instruction manuals, photos, videos of operations, maintenance history, and manufacturer information. Furthermore, a 3D BIM model improves upon the traditional 2D drawings that had difficulty illustrating the vertical location or position of facilities.

The BIMFM system consists of subsystems for BIM, 2D barcodes, mobile devices, and a hub center. The BIM, 2D barcodes, and mobile devices subsystems are located on the client side, while the hub center subsystem is on the server side. Each subsystem is briefly described below.

## BIM Subsystem of the BIMFM System

In this study, BIM is used as an information model in the BIMFM system and applied to capture and store information about the facility, including basic descriptions, parameter-related information, maintenance records, and interface reports. Autodesk Revit software was used to create the BIM model files. Autodesk Design Review was used to read the BIM models of facilities. Information integration with the 3D BIM models was achieved using the Autodesk Revit application programming interface (API) and the Microsoft Visual Basic.Net (VB. Net) programming language. The BIMFM system was developed by integrating the 3D BIM models of facilities and maintenance-related information using Revit API programming. ODBC was utilized to integrate the acquired data from different software programs and all maintenance information, such that BIM files, can be exported to an ODBC database for connection with the BIMFM system.

## 2D Barcode Subsystem of the BIMFM System

Most people in Taiwan have personal smart phones and tablets and can easily access 2D barcode information. The case study uses the QR code as the 2D barcode system since the QR code reader software is popular in Taiwan and provides the most suitable functionality for facilities maintenance management. The QR code label has a high fault tolerance and its anticorruption capability contributes to longer usage and better identification. One of the major advantages of using the 2D barcode is that no extra cost is required to buy software, since a great number of 2D barcode software applications for tablets are free. Furthermore, all types of 2D barcode labels can be created using a personal computer (PC) printer.

## Mobile Devices Subsystem of the BIMFM System

Two mobile devices are used in the BIMFM system. A Samsung Series 7 tablet is used as the webcam-enabled tablet hardware. The Samsung Galaxy Tab runs on Windows 8. All data in the tablet module are transmitted to the server directly through the Internet. An HP Pavilion notebook is used as the webcam-enabled notebook hardware. The HP Pavilion notebook runs on the Windows 7 operating system. All data in the tablet and the notebook are transmitted to the server directly through the Internet via Wi-Fi or third generation (3G).

## Hub Center Subsystem of the BIMFM System

The hub center is an information center in the BIMFM system that enables all participants to log on to a hub center and immediately obtain information required for FM. Users can access different information and services via a single front-end access point on the Internet. For example, FM staff can log on to the hub center and securely access the latest FM schedule information. FM managers can check maintenance status, results, and various other inspection-related data. All facilities-related information acquired within the hub center subsystem is recorded in a centralized system database. FM staff can access the required information via the hub center subsystem based on their access privileges.

The amount of maintenance information stored will increase over time if all FM information is recorded in the BIM model. Because BIM models cover a wealth of building information, system storage space should be reserved for crucial information, such as spatial information, facility ID and name of the facility, facility location, and other critical information. In order to keep the system performance at an acceptable level, the information derived by other applications should be stored in an external location. Therefore, two databases are incorporated into the design of the BIMFM system: the BIM elements database and the FM database. The BIM elements database stores only basic information (such as the position, ID, and name of the facility and key parameter information of components). Related maintenance data and information are stored in the FM database.

The accuracy of the BIM model will directly affect FM operations in the BIMFM system. To prevent too many users from simultaneously using the BIM models and, in turn, affecting their accuracy, the BIM engineer can update the information from the BIM elements database directly in the BIMFM system. The latest information in the BIM elements database automatically resyncs when content changes. In this framework, all building facility information from BIM can be saved and updated in the BIM elements database without directly accessing the BIM models.

FM operations do not require all building information; they only require information about necessary maintenance, although the BIM model may cover the whole building. Therefore, during the pre-FM process, the BIM engineer is responsible for determining whether to create the DWF (design web format) file of the BIM model in advance and save it as a source for decomposed BIM models based on the requirements of the FM operations. Not only can the DWF format retain building information, but also its file size is smaller than the general BIM model file. The BIMFM system can be improved with the use of the DWF file for the 3D BIM illustrations, and the system's performance is enhanced for users by reviewing the 3D BIM models. Furthermore, the BIM elements database in the server can store accurate information on the BIM models.

In the BIMFM system, the following three major roles are involved in FM: a BIM engineer, an FM manager, and FM staff. To ensure that the FM operation does not affect the maintenance operation of the BIM model, this study utilizes client-server system architecture. In the BIMFM system, the BIM elements database stores all of the information on the BIM models on the server side. In addition, only BIM engineers are allowed to access and edit the BIM models and export data to the BIM elements database using the BIM software directly on the server side. On the client side, the FM manager and FM staff refer to facility information through the BIM elements database and edit FM information through the FM database in the BIMFM system.

The BIMFM system server supports four distinct layers, each with its own responsibilities: management, data access, application, and presentation. The following section describes these four distinct layers in the BIMFM system.

The management layer provides BIM engineers with the tools to edit and manage BIM models using the BIM software. BIM engineers can access and edit the BIM models saved in the server through the Internet. With the development of the BIM tool APIs, the management layer can not only export data from the BIM models to the BIM elements database but also import data from the BIM elements database to the BIM models. Furthermore, facilities maintenance information can also be recorded in the BIM elements database in the management layer.

The database layer in the BIMFM system consists of two databases: the FM database and the BIM elements database. The FM database stores all facilities maintenance records, while the BIM elements database stores complete facility information, including facility number, name, and type, in the BIM models. The FM database records detailed maintenance information in accordance with the facility ID. The primary key establishes a relationship between the facility ID and the main index. Therefore, information can be used for data association for data mapping to retrieve complete facilities maintenance information based on the facility ID between the two databases.

The application layer defines various applications for the major system and API modules. These applications offer indexing, BIM model data updates and transfers, facility status visualization, and report generation functions. The application layer integrates and uses the BIM software to open the BIM models using developed API modules. Finally, the application layer can automatically acquire data and analyze the BIM models based on a request and then send the results back to the client side.

The presentation layer is the main implementation platform of the BIMFM system. During the FM process, the FM manager and FM staff can use a tablet (client side) and the utilities in the BIMFM system for the FM operation. The presentation layer is integrated with a QR code device, automatically displays the location information of the BIM model, records maintenance information, illustrates the different conditions and status of FM, queries the history, and exports reports on FM results.

# SYSTEM DEVELOPMENT

The BIMFM system server is based on the Microsoft Windows Server 2008 operating system with an SQL Server 2008 R2 as the database. The BIMFM system is developed using VB.NET programming, which is easily incorporated with ADO.NET to transact FM and BIM information with an SQL Server database. The BIMFM system consists of three different user areas, FM staff, FM manager, and BIM engineer areas. Access to the BIMFM system is password-controlled.

## System Functionality Description

This section describes the implementation of each major functionality module in the BIMFM system (see Figure 2).

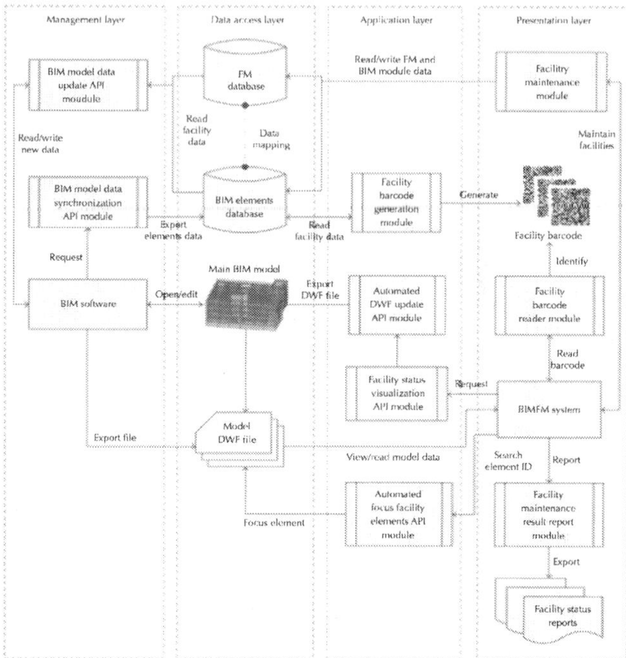


**Figure 2:** System and module framework of the BIMFM system.

## FM Information Functional Module

The functional module provides FM staff with detailed FM information on facilities by reviewing 3D BIM models. This module enables all FM staff to refer to related FM information and historical maintenance records for the selected facility quickly and easily in the 3D BIM-based environment. This module allows FM staff to refer to basic information and specifications associated with 3D BIM models during the FM process. This module also has a search function that enables the information to be found and retrieved easily.

## FM Maintenance Functional Module

FM staff can download up-to-date maintenance records through the 3D BIM models and enter facility maintenance results directly into the 3D

BIM models. Additionally, the module can automatically produce the corresponding maintenance forms through the 3D BIM models. Tablets display the checklist for every facility maintenance task. FM staff can record maintenance information such as dates, conditions, inspection results, descriptions of problems that have arisen during maintenance, and recommendations. Furthermore, FM staff can also check tasks that do not pass the inspection and select relevant tasks from lists in the 3D BIM models. One of the benefits of the module is that maintenance results and records can be transferred between a tablet and the BIMFM system by real-time synchronization, eliminating the need to enter the same data more than once.

## FM Process Monitor Functional Module

This functional module is designed to enable FM managers to monitor the FM process. The process monitor module provides an easily accessed and portable environment where FM staff can trace and record all maintenance information and status through the visualized and colorized BIM model.

## FM Reports Functional Module

Users can easily access the FM reports functional module to identify needs and analyze FM results information. Authorized records for interfaces can be extracted and summarized for the final FM result-related reports. Furthermore, all FM reports can be extracted using commercially available software such as Microsoft Excel.

## System API Modules Description

In order to integrate the system with the BIM models, the following API modules are developed in the BIMFM system.

## The BIM Model Data Synchronization API Module

The main function of this module is to automatically synchronize the latest information on the BIM models with the BIM elements database.

Although the BIM software provides the open database connectivity (ODBC) database export function, there are still many required data elements for FM that cannot be exported through this function, such as self-defined parameters information in the BIM elements modules. That functionality is provided by the API. All required information in the BIM models is automatically synchronized to the BIM elements database based on required information for FM by the API development. The module will retain the existing data and update the changed data synchronization if the exported information already exists in the BIM elements database.

## *Facility Barcode Generation and Reader API Module*

This module generates a QR code label automatically and links to the related facility BIM module or BIM element module. Because there are typically thousands of facilities for FM, the FM staff usually requires significant time and effort to make QR code labels for FM purposes. To address this problem, the module can generate the QR code label to obtain basic required information (such as device number and purchase date) and linkage to related 3D BIM modules. By scanning the QR code, users can link and check the related BIM modules directly without spending too much time searching for them. Furthermore, a QR code can be generated and accessed quickly to find available BIM models in the room corresponding to the facility location.

## *Automated Focus on Facility Elements API Module*

This module allows users to access the related BIM models by scanning a QR code label attached to the surface of the facility. When the user scans the QR code label, this module will automatically identify the facility or facility location based on the corresponding floor to open the corresponding BIM models file (in DWF format). If the facility is positioned too high up to scan its QR code label, the user may scan the QR code label attached to the entrance of the room and directly select and access the corresponding BIM models in the room.

## BIM Model Data Update API Module

This module provides the functionality for updating facility information from the FM database to the main BIM model automatically. The DWF format of BIM models in the BIMFM system allows users to view and access BIM models without changing any information in the BIM models. When FM-related information changes and requires feedback to the BIM model (e.g., the facility is lost or scrapped), the most recent maintenance date and the facility replacement date can be automatically updated for the corresponding BIM model. Therefore, BIM engineers and FM managers can directly access the updated facility maintenance information in BIM software.

## Automated Updated DWF File API Module

This module is mainly to allow users to quickly access the latest BIM models in the BIMFM system through the updated whole or separated DWF file. When the size of BIM models increases, system performance slows down. A solution is to decompose whole BIM models into smaller BIM models (exported as separate DWF files) for improved system performance. Models may be decomposed according to a floor or a specific area. Furthermore, the module will update the separated DWF file automatically when any information changes in the BIM model.

## Facility Status Visualization API Module

The module provides the visualization functionality for FM status through a visualized BIM model. Through a systematic FM analysis of test results, the module displays different colors to illustrate various conditions and FM status (such as qualified inspection, required repair status, and obsolete facility). Users can access the overall different maintenance conditions and FM status quickly through the visualized BIM model. Table 1displays the colors associated with each status.

**Table 1:** Description of color usage in BIM model

| Color usage | Description |
|---|---|
| Green | The facility's maintenance work has been completed and the result is satisfactory. |
| Red | The facility's maintenance problem has been identified, but the result is not satisfactory. |
| Yellow | The facility's maintenance work has been out of schedule and the facility's maintenance work has not yet started. |
| Blue | The facility's maintenance work has been out of schedule and the facility's maintenance work has not been completed. |

There are two subsystems in the BIMFM system. The first subsystem is the API monitoring subsystem for BIM engineers located on the server side. This subsystem deals with integration services of BIM models in the BIMFM system. These services include BIM elements database initialization, updating facility maintenance information, and visualizing the maintenance status of facilities. Another subsystem is the maintenance subsystem located on the client side. This maintenance subsystem is developed for FM staff and FM managers to deal with FM operations in the facility's location, such as reading the barcode attached to the facility, recording FM, and reporting FM results.

## System Process Description

There are four processes used in the BIMFM System including the system initialization process, FM information monitoring process, maintenance implementation process, and API information processing process (see Figure 3).

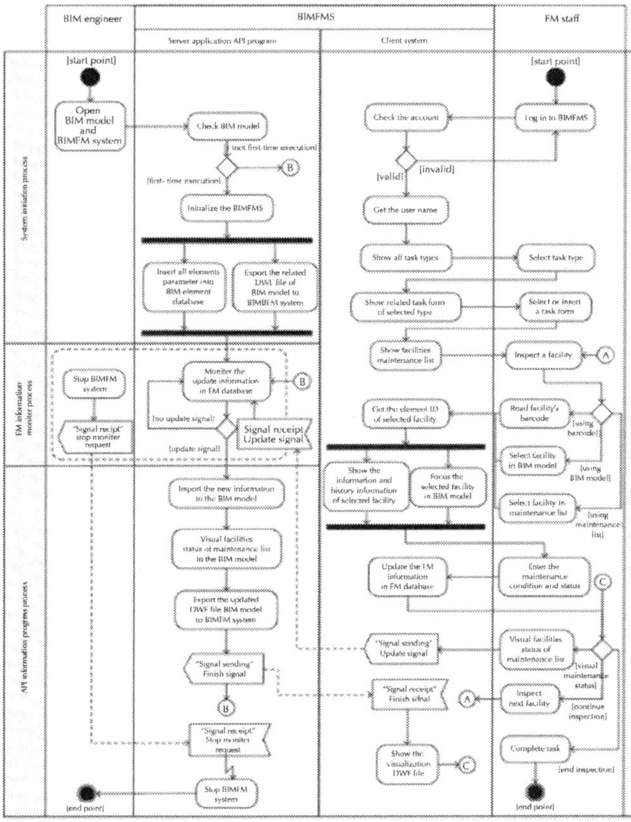


**Figure 3:** The system process flowchart used in the BIMFM system.

## System Initialization Process

The purpose of the system initialization process is to provide adequate information on a facility for maintenance operations. The BIM model must provide all information and related models (DWF files) on a facility as an information requirement for facility maintenance operations. When the BIM model is input with complete facility information, the BIM engineer needs only to use BIM software (such as Revit) to open the BIM model and run the BIMFM API monitoring system to complete the setup work. When the BIM engineer opens the API monitoring system, the system will automatically determine whether the BIM model is run in the program for the first time. If so, the system will automatically insert all the facility elements information into the BIM elements database and the BIM 3D model (such as the model of each floor and a special area model) for exporting to the DWF file in the BIMFM system. If not, the system will only automatically update any new facility elements information to the BIM elements database and update the changed DWF file.

## FM Information Monitoring Process

When the system initialization process is completed, the system will automatically enter the FM information monitoring process. The major purpose of the FM information monitoring process is to check and track whether the user requests the server to update API information. When the demand signal is transmitted to the system by a user, the system will begin to update API information again in the server. Furthermore, the BIM engineer can stop the FM information monitoring process at any time. All API services in the application layer of the BIMFM system will stop operation when the FM information monitoring process is stopped.

## Maintenance Implementation Process

During the maintenance implementation process, the maintenance list varies according to the maintenance task categories. The design lets FM staff work on maintenance operations effectively according to the task categories and maintenance list. FM staff can utilize the webcam-enabled Tablet PC to access the BIMFM system and show all the task categories and maintenance lists based on different levels of access. After FM staff selects a particular task category, the system shows the history task form for that category. FM staff can view the other task forms, edit the unfinished task form, or add a new task form. When FM staff selects or adds a task form, the system retrieves facility information from the BIM elements database based on the task types. Furthermore, a list of all related maintenance and results will be illustrated with the BIM model for FM work preparation. FM staff can access inspection information and the maintenance status effectively. During the maintenance implementation process, FM staff can use the system directly and read the QR code attached to the surface of the facility. When the system receives the facility ID, the system automatically displays the facility's basic information and historical maintenance data in the BIM model. Furthermore, the facility's BIM model will be selected, focused, and highlighted using different color. User can obtain basic information on the facility by scanning the QR code attached to the facility, clicking the BIM model, or selecting from a maintenance list. After selecting the facility through one of the three methods, the FM staff can handle maintenance work and record the status and result of maintenance. Finally, all maintenance records and information are stored in the FM database.

## API Information Processing Process

During the process of maintenance operations, the maintenance status can be enhanced by color visualization in the BIM model through the API information process. Through the functionality that visually depicts the status of maintenance list items, the system will send the updated signal automatically to the server side of the BIMFM system. The system will start API information processing if the BIMFM system is running during the FM information monitoring process. First, the system will get related maintenance information from the maintenance list in the FM database and update the main BIM Model through API information processing. After the maintenance list is updated in the database, API creates a new 3D view automatically; the 3D view assists with color visualization of the facility based on the maintenance status. The 3D BIM model of facilities in the selected task form will be displayed in a color based on maintenance status, while the rest of the BIM model elements will be displayed in translucent white to enhance the visualization effect. All color visualization is described in Table 1. Finally, the system will export the completed 3D view BIM model to the DWF format automatically, store DWF files in the server side, and return the completed signal to the client side. The system will automatically connect to the server and open the visual DWF files to assist in the visual effect of the maintenance status when the client receives the completion signal.

## System Information Flow Description

There are three external entities in the BIMFM system. They are entity BIM engineer, entity FM staff, and entity FM manager (see Figure 4). This section describes each external entity in the BIMFM system.

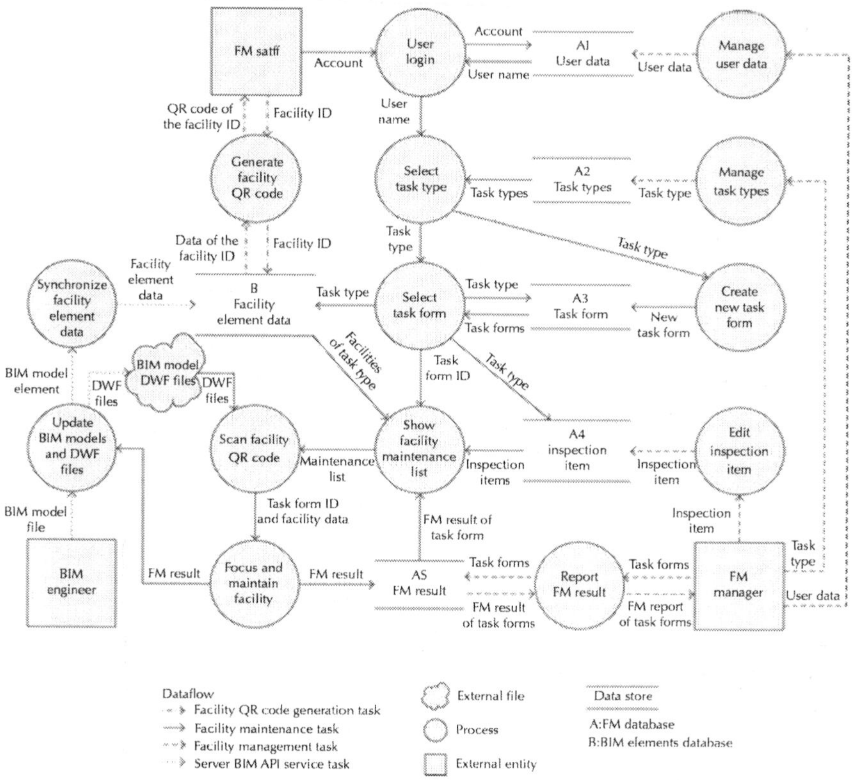


**Figure 4:** The information flow used in the BIMFM system.

The entity BIM engineer is primarily responsible for starting the BIMFM system in the BIM API service on the server side and opening the applied BIM model in the system. This is called the BIM API service task in the study. In the beginning, entity BIM engineer must use the BIM software to open the BIM model and also start the BIM API service.

When the BIM model is imported to the BIMFM system, the process update BIM models and DWF files will automatically update the BIM model information (including BIM model elements and the DWF files). The process synchronize facility elements data will check the updated facility element data first. The needed updated information will be imported into the store facility elements data in the BIM elements database. All updated DWF files will be saved directly in an external file.

The entity FM manager is primarily responsible for handling the BIMFM planning and management work. The operation is called the FM task in the study. The FM tasks include various operations and procedures of management (including processes such as manage user data, manage task types, and edit inspection item). The entity FM manager can handle all system data management during the processes. All changed and updated information will be imported into the corresponding data store of the FM database. Furthermore, the entityFM manager can select multitask forms to export as a report. The process report FM result will acquire the related information based on selected task forms from table FM result data of the FM database. Furthermore, the process report FM result analyzes the information, compiles it into report format, and sends back a final report to the entity FM manager.

The entity FM staff is primarily responsible for handling the facility QR code generation task and facility maintenance task. The facility QR code generation task is developed to handle the preparation of QR code work. Entity FM staff may create one QR code label or a set of QR code labels for FM use. The process generate facility QR code will obtain related information on the facility from the BIM elements data table based on the facility ID, send the information for QR code coding, create the QR code image file for the facility, and send it back to entity FM staff. The facility maintenance task handles various operations and procedures of FM. In the beginning, the process user login will check the user's authority based on the user name and password in the store user data. The process select task type will show task types in the store task types for the selection if certification is passed. The process select task form will show the related FM task form based on task type for FM staff selected. If the maintenance work is a new activity, the process create new task form will create a new task form based on the selected task type and save it in the table task form data. After entity FM staff selects a task form, the process select task form will acquire the necessary facilities list from the table facility elements data in the BIM element database and acquire necessary inspection items from table inspection item data in the FM database. The facilities list and inspection items integrated with the selected task form ID will be imported into the show process facility maintenance list. Furthermore, the show process facility maintenance list acquires maintenance results from table FM result data in the FM database. The process show facility maintenance listwill arrange information as maintenance list and export information to the process scan facility QR code after acquiring complete information. When entity FM staff scans the QR code of the facility, the facility QR code process will decode the QR code automatically to send task form ID and facility data into the process focus and maintain facility.

The process focus and maintain facility will zoom in and highlight the BIM model of the facility automatically when it receives the facility data. Final maintenance results will be updated in the table FM result data in the FM database when entity FM staff finishes the maintenance work and sends back the process update BIM models to update facility status in main BIM model. The process update DWF files in the BIM API service will update all changed DWF files automatically, save them in the external file, and send DWF files back to the process scan facility QR code to let FM staff review the visualized DWF file if FM staff requests facility status visualization.

Integrated with the above design concept, more complex operating procedures of FM are simplified and developed in the BIMFM system. One of the major characteristics of the BIMFM system is to provide users with easy-to-use visualization for handling FM work. By clicking the list, each task form will show the list of facilities requiring maintenance, historical maintenance information, and the status and condition of facilities maintenance. By scanning the QR code attached to the facility, the corresponding BIM models are linked and illustrated quickly and effectively in facility location. Finally, all maintenance results are sent back and saved in the main BIM model. The proposed approach provides a means to update the facility information of the BIM model and FM information synchronization. Finally, in order to let FM staff applies the system easily and effectively, the layout of the system is designed based on FM staff's suggestions. Figure 5 shows the graphical user interface (GUI) of the BIMFM system.

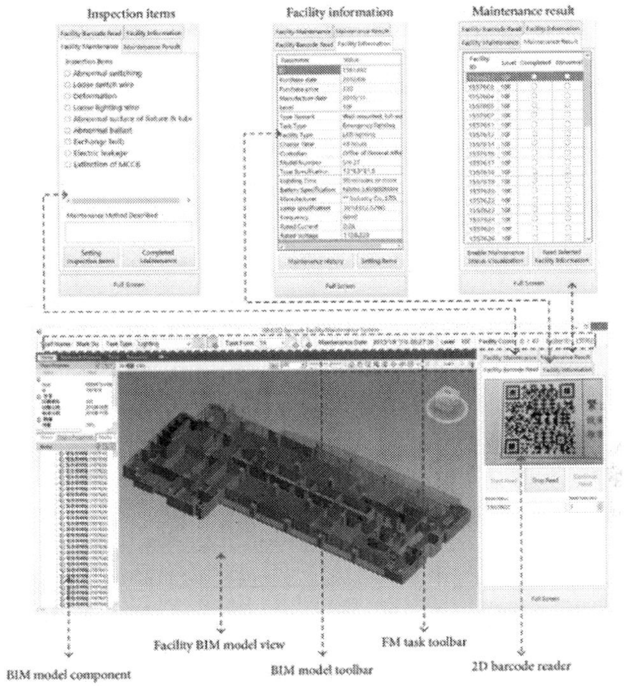


**Figure 5:** GUI of the BIMFM system.

# SYSTEM VALIDATION

## Case Study

For the case study, the proposed BIMFM system of this study was applied to a building in Taiwan; that is, a BIMFM system was utilized for the FM of the case study building, which contained approximately 20 facilities that had to be managed and inspected. Usually, the FM work was executed every month. Existing approaches for tracking and managing the FM work relied on paper-based records. The bulk of the FM work was paper-based and documented by repeated manual entry, although an FM system was developed for a standalone software application. Therefore, the FM staff in the FM division utilized the

BIMFM system to enhance the FM work in the case study.

After the critical facilities were selected for FM work, each QR code label was made, and the unique ID for each facility was entered into the BIMFM system database for quick search. During the FM process, the QR code label was scanned for basic information about the facility before the FM work started. Before the FM work began, the FM staff could check the facility list from webcam-enabled tablets, refer to the relevant information, and begin preparation work without printing any paper documents. During the FM process, the FM staff scanned the QR code label first (see Figure 6(a)). The BIMFM system showed the basic information and the BIM model of the facility after scanning the QR code label. The FM staff could then check further detailed information like maintenance instructions, notifications, and accessories list, all of which were supported by BIMFM (see Figure 6(b)). After the FM work, the FM staff entered the results of maintenance, edited the description in the tablet, and provided the updated information to the system (see Figure 6(c)). When a facility required repairs, the system also provided the manufacturer's problem information for immediate reference. Finally, the facilities manager and the authorized FM staff accessed the updated information simultaneously from their offices (see Figure 6(d)).

(a) FM staff scanned the QR code and accessed BIM model of facility

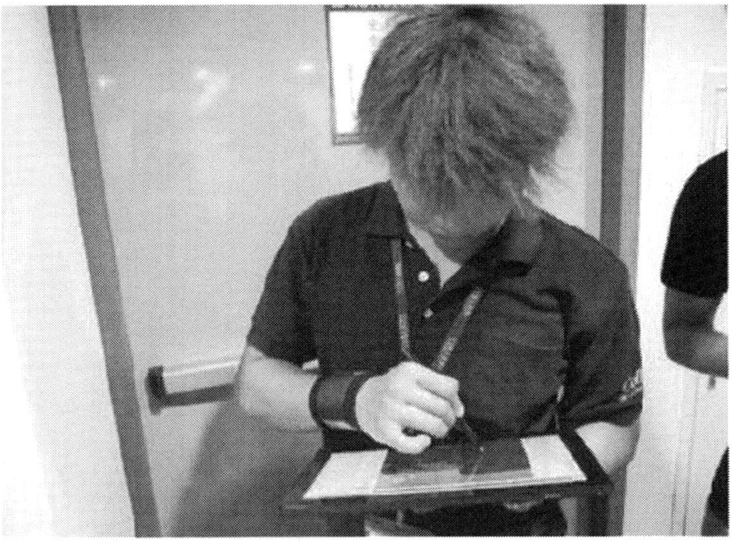

(b) FM staff referred to the maintenance list and reviewed the BIM model in the tablet

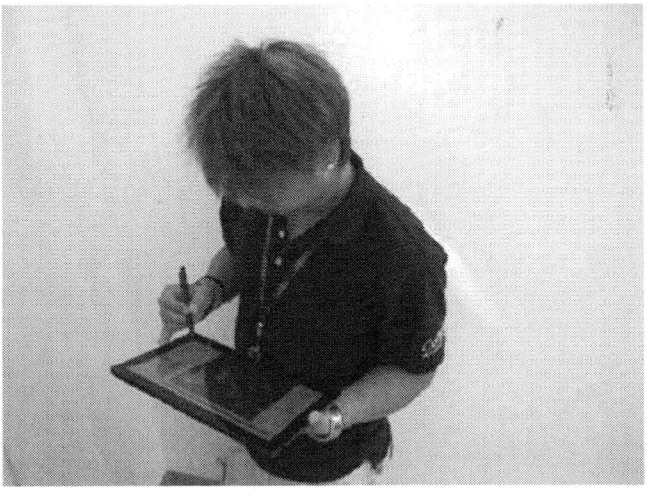

(c) FM staff inputted and updated the maintenance records in the tablet

(d) FM manager accessed maintenance records directly using BIMFM system.

**Figure 6:** FM staff using webcam-enabled tablet to scan QR code label for FM work in the case study.

## Evaluation and Results

Overall, the field test results indicate that the integration of BIM and 2D barcode labels is an effective tool for the FM of a building. All 2D barcode labels survived use in the pilot test over the two-month testing period. Approximately 25 users participated in field trials of the FM process. The BIMFM system was installed on the main server in the FM division of the building.

During the field trials, verification and validation tests were performed to evaluate the system. The verification test aims to evaluate whether the system operates correctly according to the design and specification, while the validation test assesses the usefulness of the system. The verification test was carried out by checking whether the BIMFM system could perform the tasks specified in the system analysis and design. The validation test was undertaken by asking selected case participants to use the system and provide feedback by answering a questionnaire. Twenty-five participants were involved in the evaluation test. To evaluate the system function and the level of satisfaction with the system's capabilities, the users of the system were asked to grade the conditions of system testing, system function, and system capability, separately, in comparison with the typical paper-based FM approach. Some comments for future improvements to the BIMFM system were also obtained from the case participants through the user satisfaction survey. Finally, Table 2 shows a comparison of the current approach and the proposed system.

**Table 2:** Comparison of current approach and proposed system

| Item | Current approach | | Proposed approach | |
|---|---|---|---|---|
| | Method | Average time | Method | Average time |
| Edit the defect problems of the facility | Edit the defect problems by paper-based sheet | 12–20 sec | Edit the defect problems through the BIMFM system | 6–12 sec |
| Find basic information on facility for reference | Review maintenance data on paper-based sheet | 12–23 sec | Access basic information on a facility directly by accessing and clicking BIM model | 7–13 sec |
| Refer to relevant historical maintenance information | Refer to paper-based maintenance lists and reports | 12–18 sec | Click BIM model and refer to historical maintenance information directly | 6–12 sec |
| Maintenance information updating | Identify and record results on a checklist, then re-enter at the office | 42–52 sec | Real-time data entry in the system during maintenance process | 22–42 sec |
| Mark the inspection problems at the facility location | Refer the paper-based maintenance condition and status sheet | 1–1.5 min | Illustrate overall maintenance conditions and status of FM quickly through visualized BIM model | 40–50 sec |

The percentage of satisfied users (96%) obtained from the user satisfaction survey indicates that the BIMFM system is quite adaptable to current FM practices in a building and is attractive to users. The overall result implies that the BIMFM system is considered well designed and is able to enhance current time-consuming FM processes. The satisfaction rate exceeding 88% also indicates that the visual BIM model that provides FM support is very helpful. The 92% rate of satisfaction with the integration of the QR code in the BIMFM system for the direct access of the BIM model also indicates that this integration of the QR code is considered effective and necessary. Moreover, no additional work is required to complete the documentation beyond the data collection process. The advantages and disadvantages of the BIMFM system identified from the pilot study are identified.

In the cost analysis, the total cost of the equipment applied in this study was $3,500 US dollars (including an 11-inch webcam-enabled tablet and one PC server). Most personal computers can generate and print QR code labels using free software. Furthermore, there is no additional cost for the QR code reader hardware because most tablets are equipped with cameras that enable 2D barcode scanning. The experimental results demonstrate that the BIMFM system can enhance the visual FM process significantly and effectively when using a BIM approach that is integrated with 2D barcode technology. The use of these technologies significantly improves the overall performance of maintenance operations.

## Limitations and Barriers

The findings of this case study revealed several limitations of the BIMFM system. The following are inherent problems recognized during the case study.(i)It is difficult for new users to operate the BIM model in the BIMFM system. Some FM staff are initially unfamiliar with BIM models. As it usually takes time to learn how to use BIM models, the use of the BIM system in the case study initially lengthened the FM operation over the traditional approach, since users required time to find the corresponding BIM model and fill out the FM information in the BIMFM system. After the user becomes skilled and familiar with the BIM model, the time required by the current approach and the proposed system becomes almost exactly the same as in the previous FM operations.(ii)If BIM models do not exist for the purpose of construction management during the construction phase, the BIM approach integrated with FM will not likely be implemented within the BIM environment. Most FM companies do not want to spend the required time and cost to use BIM for only FM work on building projects.(iii)As QR code labels attached to outdoor facilities are easily damaged because of external environmental pollution (such as dust and rain), it is necessary to consider and enhance the protection and the waterproofing of the QR code labels.(iv)The QR code technology's short-read distance range was the primary limitation in the case study. In some facilities, the QR code labels were installed up high where FM staff could not easily reach them. Therefore, it is recommended that the QR code label be attached at a lower space on the facility for easier scanning or, alternatively, that the QR code label be associated with the BIM models for all of the facilities in the room and placed at the entrance of the room. A user can then scan the QR code label at the entrance of the room to find and select directly the BIM model of a facility.(v)The best read distance of QR code labels is about 3 meters (in a straight line). Based on the case study, the scanning distance varies depending on the tablet camera's resolution. Furthermore, the webcam-enabled tablet cannot identify a QR code label and read the information if the lighting is too low. The QR code labels cannot be recognized if the corner side of the position-detection pattern block position is damaged or polluted.(vi)In consideration of the limited storage capacity of the tablet and the notebook, it is suggested that the BIM engineer create the DWF files in the database in advance. However, the file size of the DWF will affect the performance of the BIMFM system directly and evidently. The impact includes the time of reading of the DWF files, the time to search for the facilities, and the smoothness of the system operation. When the DWF file is too large (more than 100 MB), it cannot be opened. Therefore, the main BIM model of the whole building should be exported and separated into many DWF files for each floor. When the FM staff execute the FM work, the system opens the DWF file for the floor only after the FM staff scan the QR code for the facility. Furthermore, the FM staff can quickly refer to the separated DWF files of the floor for FM work.(vii)Based on the case study, BIM engineers are required to update the BIM models continuously during the maintenance and operation phase. When new equipment or facilities are purchased, the BIM engineers must create an FM element for the new equipment in the BIM model for future maintenance use. Furthermore, communications between the FM maintenance staff and the BIM engineers are necessary and important during the process. The FM staff should inform the BIM engineers about any problems regarding the BIM models. After the BIM engineers correct the BIM models, they must also notify and discuss it with the FM staff. The BIM models require constant maintenance and updates. Another important issue is the quality management of the BIM models. Although the study proposes using the BIMFM system as a means of helping FM staff handle visual facilities maintenance and management work, the advanced management procedures and mechanisms for the quality management of the BIM models for FM must be identified and developed in the future. Particularly, the management mechanisms for updating the BIM models should be developed as the next step of BIMFM system development.

# CONCLUSIONS

The BIM approach, which is applied to retain facility information in a digital format, facilitates easy updating of FM information in a BIM environment. Although many practical cases of using BIM during the maintenance management stage exist, one problem is that it is not easy for FM staff to find the corresponding FM element in the BIM model for maintenance management during the phase. In order to assist FM staff with obtaining the corresponding FM element in the BIM model for FM in an automatic and effective manner, this study develops a BIMFM system that integrates 2D barcode technology to connect automatically to the BIM models. The mobile automated BIMFM system not only improves FM efficiency but also provides a real-time service platform during the FM process. In the case study, 2D barcode readings increased the accuracy and the speed of the BIM model searches, indirectly enhancing performance and productivity. The FM staff used webcam-enabled tablets to enhance the FM work seamlessly at facility locations, owing to the system's searching speed and ability to support related information collection and access during the FM process. Meanwhile, on the server side, the mobile automated BIMFM system offers a hub center to provide the FM division with real-time monitoring capacity during the FM process. Integrated with the characteristics of 3D BIM model illustration and BIM parametric design, the mobile automated BIMFM system quickly shows the necessary maintenance information by using a facility's BIM model based on the selected task type and clearly presents the position and the height of the selected facility. The main contribution of this study is to help FM staff obtain the corresponding FM element in the BIM model in an automatic and effective manner by integrating BIM with 2D barcode technology. Furthermore, the proposed solution aims to enhance the tracing and recording of FM status through the visualized and colorized BIM model.

In the case study, the application of the mobile automated BIMFM system helped to improve the FM work of a commercial building in Taiwan. Based on the experimental results, this study demonstrated that BIM technology has the significant potential to enhance FM work. The integration of BIM technology with 2D barcode technology helps FM managers and FM staff to effectively track and control the whole FM process. Compared with current approaches, the combined results demonstrate that a BIMFM system can be a useful mobile BIM/2D barcode-based FM platform. Based on the case study findings, the BIM models must be constantly updated and corrected. Another important issue is the quality management of the BIM models. The advanced management procedures and mechanisms for the quality management of the BIM models for FM need to be identified and developed in the future. Such an endeavor will be the next step in BIMFM system development. Finally, the limitations, encountered problems, and suggestions are discussed based on the implementation of the case studies in this study. Despite the challenges indicated above, the promising results shown in this study demonstrate the great potential of the proposed system as a means of aiding the FM of buildings.

## CONFLICT OF INTERESTS

The authors declare that there is no conflict of interests regarding the publication of this paper.

## ACKNOWLEDGMENTS

The authors would like to express their appreciation to S-W Property Company for assistance in the system verification and validation and to the staff and managers for providing useful data, valuable information, and helpful comments during system test.

# REFERENCES


1. S. Boddy, Y. Rezgui, G. Cooper, and M. Wetherill, "Computer integrated construction: a review and proposals for future direction," Advances in Engineering Software, vol. 38, no. 10, pp. 677–687, 2007.
2. N. Li and B. Becerik-Gerber, "Performance-based evaluation of RFID-based indoor location sensing solutions for the built environment," Advanced Engineering Informatics, vol. 25, no. 3, pp. 535–546, 2011.
3. R. Manning and J. Messner, "Case studies in BIM implementation for programming of healthcare facilities," ITCON, vol. 13, pp. 446–457, 2008,
4. Y. C. Lin, "Challenges in BIM implementation for FM," in Proceedings of the Project Management and Property Management Conference, p. 6, Taipei, Taiwan, November 2011.
5. L. E. Bernold, "Testing bar-code technology in construction environment," Journal of Construction Engineering and Management, vol. 116, no. 4, pp. 643–655, 1989.
6. L. C. Bell and B. G. McCullouch, "Bar code applications in construction," Journal of Construction Engineering and Management, vol. 114, no. 2, pp. 263–278, 1988.
7. L. H. Blakey, "Bar codes: prescription for precision, performance, and productivity," Journal of Construction Engineering and Management, vol. 116, no. 3, pp. 468–479, 1990.
8. G. Stukhart and E. L. Cook, "Bar-code standardization in industrial construction," Journal of Construction Engineering and Management, vol. 116, no. 3, pp. 416–431, 1990.
9. L. E. Bernold, "Bar code-driven equipment and materials tracking for construction," Journal of Computing in Civil Engineering, vol. 4, no. 4, pp. 381–395, 1990.
10. B. G. McCullouch and K. Lueprasert, "2D bar-code applications in construction," Journal of Construction Engineering and Management, vol. 120, no. 4, pp. 739–752, 1994.
11. R. Navon and O. Berkovich, "Development and on-site evaluation of an automated materials management and control model,"

Journal of Construction Engineering and Management, vol. 131, no. 12, pp. 1328–1336, 2005.

12. T. Shehab and O. Moselhi, "An automated barcode system for tracking and control of engineering deliverables," in Proceeding of Construction Research Congress, San Diego, Calif, USA, April 2005.
13. G. Saeed, A. Brown, M. Knight, and M. Winchester, "Delivery of pedestrian real-time location and routing information to mobile architectural guide," Automation in Construction, vol. 19, no. 4, pp. 502–517, 2010.
14. H. Li and W. W. S. Lu, "Building information modeling and changing construction practices,"Automation in Construction, vol. 20, no. 2, pp. 99–100, 2011.
15. T. K. Tse, K. A. Wong, and K. F. Wong, "The utilisation of building information models in nD modelling: a study of data interfacing and adoption barriers," Electronic Journal of Information Technology in Construction, vol. 10, pp. 85–110, 2005.
16. C. Eastman, P. Teicholz, R. Sacks, and K. Liston, BIM Handbook: A Guide to Building Information Modeling for Owners, Managers, Designers, Engineers and Contractors, Wiley, Hoboken, NJ, USA, 2008.
17. Z. Shen and R. Issa, "Quantitative evaluation of the BIM assisted construction detailed cost estimates,"ITCON, vol. 15, pp. 234–257, 2010, B. Succar, "Building information modelling framework: a research and delivery foundation for industry stakeholders," Automation in Construction, vol. 18, no. 3, pp. 357–375, 2009.
18. B. Becerik-Gerber and S. Rice, "The perceived value of building information modeling in the US building industry," ITCON, vol. 15, pp. 185–201, 2010,
19. Y. Jung and M. Joo, "Building information modelling (BIM) framework for practical implementation,"Automation in Construction, vol. 20, no. 2, pp. 126–133, 2011.
20. K. Barlish and K. Sullivan, "How to measure the benefits of BIM—a case study approach," Automation in Construction, vol. 24, pp. 149–159, 2012.

21. B. Hardin, BIM and Construction Management: Proven Tools, Methods, and Workflows, John Wiley & Sons, Hoboken, NJ, USA, 2009.
22. S. Staub-French and A. Khanzode, "3D and 4D modeling for design and construction coordination: issues and lessons learned," ITCON, 12, 381–407, 2007,
23. J. Underwood and U. Isikdag, Handbook of Research on Building Information Modeling and Construction Informatics: Concepts and Technologies, IGI Global, 2010.
24. CRC for Construction Innovation, "Adopting BIM for facilities management: solutions for managing the Sydney opera house," 2010,
25. A. Motamedi, A. Hammad, and Y. Asen, "Knowledge-assisted BIM-based visual analytics for failure root cause detection in facilities management," Automation in Construction, vol. 43, pp. 73–83, 2014.
26. B. Becerik-Gerber, F. Jazizadeh, N. Li, and G. Calis, "Application areas and data requirements for BIM-enabled facilities management," Journal of Construction Engineering and Management, vol. 138, no. 3, pp. 431–442, 2012.
27. Y. Wang, X. Wang, J. Wang, P. Yung, and G. Jun, "Engagement of facilities management in design stage through BIM: framework and a case study," Advances in Civil Engineering, vol. 2013, Article ID 189105, 8 pages, 2013.
28. Y. H. Lin, Y. S. Liu, G. Gao, X. G. Han, C. Y. Lai, and M. Gu, "The IFC-based path planning for 3D indoor spaces," Advanced Engineering Informatics, vol. 27, no. 2, pp. 189–205, 2013.
29. A. Costin, N. Pradhananga, and J. Teizer, "Passive RFID and BIM for real-time visualization and location tracking," in Proceedings of the Construction Research Congress, pp. 169–178, Atlanta, Ga, USA, 2014.
30. M. Gheisari, S. Goodman, J. Schmidt, G. Williams, and J. Irizarry, "Exploring BIM and mobile augmented reality use in facilities management," in Proceedings of the Construction Research Congress, pp. 1941–1950, Atlanta, Ga, USA, 2014.
31. A. Akcamete, X. Liu, B. Akinci, and J. H. Garrett, "Integrating and visualizing maintenance and repair work orders in BIM:

lessons learned from a prototype," in Proceedings of the 11th International Conference on Construction Applications of Virtual Reality, pp. 639–649, Weimar, Germany, 2011.

32. S. Azhar, "Building information modeling (BIM): trends, benefits, risks, and challenges for the AEC industry," Leadership and Management in Engineering, vol. 11, no. 3, pp. 241–252, 2011.
33. L. Y. Ding, Y. Zhou, H. B. Luo, and X. G. Wu, "Using nD technology to develop an integrated construction management system for city rail transit construction," Automation in Construction, vol. 21, no. 7, pp. 64–73, 2012.
34. C. Fu, G. Aouad, A. Lee, A. Mashall-Ponting, and S. Wu, "IFC model viewer to support nD model application," Automation in Construction, vol. 15, no. 2, pp. 178–185, 2006.
35. J. Whyte, N. Bouchlaghem, A. Thorpe, and R. McCaffer, "A survey of CAD and virtual reality within the house building industry," Engineering, Construction and Architectural Management, vol. 6, no. 4, pp. 371–379, 1999.
36. J. Whyte, N. Bouchlaghem, and A. Thorpe, "IT implementation in the construction organization,"Engineering, Construction and Architectural Management, vol. 9, no. 5-6, pp. 371–377, 2002.
37. K. W. Chau, M. Anson, and J. P. Zhang, "Four-dimensional visualization of construction scheduling and site utilization," Journal of Construction Engineering and Management, vol. 130, no. 4, pp. 598–606, 2004.
38. X. Zeng and J. Tan, "Building information modeling based on intelligent parametric technology,"Frontiers of Architecture and Civil Engineering in China, vol. 1, no. 3, pp. 367–370, 2007.
39. K. Eddy and N. Bradley, Green BIM: Successful Sustainable Design with Building Information Modeling, John Wiley & Sons, Hoboken, NJ, USA, 2008.

Chapter 9

# Simulation-Based Fuzzy Logic Approach Toassessing the Effect of Project Quality Management on Construction Performance

Gilberto A. Corona-Suárez,[1] Simaan M. AbouRizk,[2] and Stanislav Karapetrovic[3]

[1]Facultad de Ingeniería, Universidad Autónoma de Yucatán, Apartado Postal 150 Cordemex, 97111 Mérida, YUC, Mexico

[2]Department of Civil and Environmental Engineering, Hole School of Construction Engineering, 3-014 Markin/CNRL, NREF, University of Alberta, Edmonton, AB, Canada T6G 2W2

[3]Department of Mechanical Engineering, University of Alberta, Edmonton, AB, Canada

## ABSTRACT

This paper reports the development of an approach to integrate the appropriate modeling techniques for estimating the effect of project quality management (PQM) on constructionperformance. This modeling approach features a causal structure that depicts the interaction among the PQM factors affecting quality performance in a given construction operation. In addition, it makes use of fuzzy sets and fuzzy logic in order to incorporate the subjectivity and uncertainty implicit in the performance assessment of these PQM factors to discrete-event simulation models. The outcome is a simulation approach that allows experimenting with different performance levels of the PQM practices implemented in a construction project and obtaining the corresponding productivity estimates of the construction operations. These estimates are intended to facilitate the decision making regarding the improvement of a PQM system implemented in a construction project. A case study is used to demonstrate the usefulness of the proposed simulation approach for evaluating diverse performance improvement alternatives for a PQM system.

# INTRODUCTION

Discrete-event simulation has been proven to be an effective technique to deal with the uncertainty involved in the planning of construction projects [1]; project planning processes, such as scheduling [1–4], cost estimating [5, 6], risk management [7], safety management [8, 9], lean construction analysis [10–13], constructability review [14], construction logistics analysis [15–17], contractor selection [18], and productivity estimation [19–21], among other efforts, have been addressed using this technique. Moreover, several simulation models have been proposed to estimate and improve the productivity of construction operations such as pile construction [19, 22], steel fabrication [21, 23], drainage maintenance [10], pavement construction [24, 25], pipe fabrication [13, 14], pipeline construction [26], bridge construction [27, 28], concrete production [29], high-rise structural work [30], construction of concrete structures [31], and tunnel construction [32–34], among several others. Although all these models have included the effect of a number of factors affecting the productivity of these construction operations—such as weather conditions, labor experience, contractor experience, equipment condition, site conditions, and several others—they have had limited consideration of, or even overlooked, the effect of quality performance. The reason for this may be that quality performance greatly depends on the performance of organizations' management [35–37] and the complexity of management processes should get involved in the construction of simulation models attempting to incorporate such effect [38]. If that is the case, it is necessary to highlight the role of quality management at the project level on attaining quality in construction operations [36]. Therefore, project quality management (PQM) practices deserve special attention when assessing quality performance in construction projects and should be considered in simulation models with the capacity to analyze the effect of quality performance on construction operations. Nevertheless, since discrete-event simulation is a quantitative data-intensive technique, such simulation models would require developing productivity models that include the effect of PQM practices. Productivity models use historical data to analyze the factors influencing the productivity of construction operations and to estimate the effect of such factors on the productivity ([39], as quoted by [21]). In this case, there is a challenge: objective data regarding the effect of PQM on construction productivity, which can be used for productivity modeling purposes, are usually very limited or nonexistent in construction organizations [40]. For instance, when productivity measures are available, they can only be used to compute the contractor's average past performance, which serves only as a broad guideline for estimating purposes [21]. A significant amount of uncertainty gets involved and makes it difficult to assess the relationship between the PQM performance and construction productivity. In addition, the subjectivity of the variables involved in the evaluation of such relationship represents another difficulty. For example, project managers find the assessment of the quality performance easier using qualitative terms such as poor, average, or good. Furthermore, the decision makers may not have any mathematical tools available to evaluate the combined effect of the PQM initiatives on the performance goals of construction operations. This argument draws attention to the need for a system that supports decision making with knowledge-based assessments rather than with explicit information. Therefore, this paper reports the development of a simulation modeling approach to integrate the appropriate techniques for estimating the effect of PQM on construction performance. This approach should facilitate the evaluation of the PQM practices implemented in construction projects and support the decision making regarding the improvement of construction quality performance.

# MODELING THE EFFECT OF PQM ON PRODUCTIVITY

The effect of quality on productivity has been empirically verified in service [41], manufacturing [42], and even construction processes [43]. Actually, since long time ago, Deming's philosophy emphasized the significance of quality on the performance of processes and especially on the productivity: as quality improves, productivity increases due to a reduced amount of rework, fewer mistakes, fewer delays, and better use of project requirements ([44], as quoted by [41, 45]). Since the endeavor reported in this paper focuses on the construction operational level, Deming's thinking gives meaning to the convention of using, for the purposes of this modeling approach, interruptions, low productivity periods, and reworks (i.e., disruptions) for measuring quality performance and representing the effect of PQM on the productivity of construction activities. This is also consistent with the results oriented approach to quality performance that prevails in construction, in which quality is expected to be perceived in construction operations and outputs; for instance, an increase in productivity is expected when quality performance improvements are accomplished [46].

Moreover, though several authors have empirically determined the significance of the effect of quality management practices on quality performance [47–51], other ones have reasoned on the indirect path through which such effect is accomplished [43, 46, 52, 53]. The approach reported in this paper follows the reasoning mechanism proposed by Battikha [53] to identify and analyze the root causes of nonconformities. In this case, parting from the information usually found in quality management information systems, it is acknowledged that disruptions occurring during the execution of construction activities are caused by nonconformities in the management system implemented for attaining quality performance in construction projects. Based on previous research work [36, 46, 54, 55], it is also accepted that the prevention of nonconformance events during the execution of construction projects relies on the performance of PQM practices. Therefore, Figure 1 depicts the indirect path through which PQM practices influence productivity in construction activities. This path assumes that, depending on the performance of PQM practices, nonconformance events may occur during the execution of the project, which eventually may cause disruptions that will affect productivity in construction activities. Following Battikha's [53] approach to simplify the identification of patterns indicating the occurrence of nonconformance during project execution, this modeling approach also classifies nonconformance according to the process requirements with which they are related; for instance, as illustrated in Figure 1, delayed deliveries and inaccurate and out-of-specification deliveries are nonconformance events related to material supply.

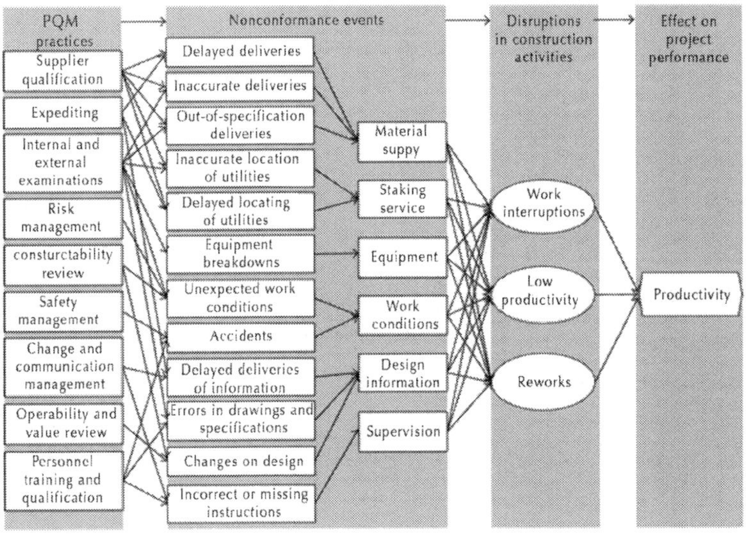


**Figure 1:** The effect of PQM practices on the productivity of construction activities.

The following sections further clarify the factors and variables involved in this model.

## Disruptions Affecting Productivity

For the purposes of this work, "disruption" is any incident that impedes a given construction activity and which, consequently, results in a negative impact on the productivity. According to the model illustrated in Figure 1, disruptions are caused by nonconformance events that occur during project execution. It is important to remark that only disruptions due to nonconformance events the project management is able to control or prevent should be accounted; for instance, disruptions due to the weather conditions should not be considered in this case.

The proposed approach intends to evaluate the consequences of disruptions in terms of the achievement of the planned productivity by means of quantifying the delays attributed to any of the following types of disruptions:

1. Interruptions, including the incidents in which an activity is temporarily discontinued due to nonconformities related either to the delays and inaccuracies associated with the supply of project requirements (i.e., labor, material, equipment, and information) or to unforeseen work conditions,
2. low-productivity periods, taking into account the times in which work progress is slower than in regular workdays; this could be due to the nonconformities concerning delays in the supply of project requirements, unforeseen work conditions, and/or poor performance of labor,
3. Reworks, denoting the cases in which nonconforming work needs to be redone in order to meet the project specifications or requirements.

The computing of the total delay in a given construction activity, which eventually will affect the productivity of the project, should include the following variables:

1. The duration of each delay due to the occurrence of any given disruption,
2. The number of disruptions occurring during the execution of the activity.

The assessment of both variables is explained later in this paper.

## Nonconformance Events Causing Disruptions

As discussed above, nonconformance events are regarded as consequences of the defective performance of PQM practices; thus they include incidents that the project managers should be able to control. They also represent the intermediate drivers that propagate the effect of the PQM practices onto the performance outcomes of construction operations (e.g., productivity).

The nature of nonconformance events that could occur during the execution of a project is usually very diverse, making it difficult to evaluate every type of them. Therefore, it makes sense to group them into categories representing the nature of a given number of nonconformance events. As said before, the proposed model has adopted Battikha's [53] approach to integrate different types of nonconformance into categories depicting requirements to accomplish activities in a given project. For instance, as illustrated in Figure 1, some of these requirements include material supply, equipment, design information, and supervision. Therefore, it is assumed that the performance of such requirements plays a major role in estimating the effect of nonconformance events on the number of disruptions and delays.

## Project Quality Management Practices Causing Nonconformance Events

The PQM practices include, for the purposes of this research, all project management initiatives that construction organizations implement to achieve quality-related objectives in construction projects. The proposed model assumes that the performance of the PQM practices is the root cause of disruptions affecting productivity during construction. For instance, Figure 1 shows that the supplier qualification practice may cause nonconformance events such as delayed deliveries and inaccurate deliveries. A number of construction PQM practices have been documented by different authors [46–48, 54, 56–58]. Figure 1 includes PQM practices that the Construction Industry Institute has proposed to assure that quality is delivered to the customers in construction projects [57]; they depict the most frequently referred types of prevention and appraisal activities in the literature [46].

The performance of PQM practices, which is the variable under evaluation in this model, depends on the maturity of the efforts applied in the implementation of such practices. In this case, the performance assessment of PQM practices involved two different measures as follows.

1. Subjective assessments based on a scale that includes five linguistic terms: there are no formal approach, reactive approach, stable formal system approach, continual improvement approach, and best-in-class performance. This approach adopted the self-assessment tool the ISO 9004:2000 standard proposes for assessing quality management processes. The interested reader should consult the ISO 9004:2000 standard for further details on this scale.
2. Objective assessments based on the calculation of a quality index that uses a crisp value scale within the 0 to 100% range. The quality index of a given PQM practice can be estimated with a self-assessment approach adopted from the model proposed by Kumar and Wolf [59] that includes a series of questions related to the level of implementation of the analyzed PQM practice. Each of these questions should be pondered with one of five possible rankings associated with a score (i.e., 0: not at all, 1: inadequate, 2: adequate, 3: good, and 4: superior) by project managers in order to obtain the sum of the scores and compute the corresponding quality index with the relation displayed as

$$\text{Quality Index} = \frac{\text{Total Rating Score}}{\text{Maximum Score}} \times 100 \qquad (1)$$

The reason to use both assessments is to reduce the effect that the ambiguity of the assessments of managers participating as developers or users of the proposed system could have on the outcomes. More objective assessments attained with the computing of quality indexes would make it possible to obtain more accurate performance estimates (i.e., productivity) of construction operations.

# SIMULATION MODELING APPROACH

The objective of this research was the development of a simulation modeling approach to assess the effect of PQM on the performance of construction operations, especially on the productivity. This endeavor implemented discrete-event and stochastic simulation in order to obtain objective productivity estimates that may facilitate the assessment of such effect. Discrete-event simulation, in which the state of the system changes discretely only at occurred events, has been the preferred approach to construction simulation because of its simplicity [26]. On the other hand, stochastic simulation, in which the state of the system changes randomly over the time based on probabilistic functions, has been regarded as especially suitable for modeling the uncertainty involved in the variable behavior of construction operations [60]. In order to facilitate the implementation of these simulation techniques, the aforementioned effect is evaluated in terms of the attainment of productivity goals in a given construction operation. This approach reduces the problem to the estimate of the probability mass functions that embody the uncertainty associated with the occurrence of disruptions that may delay the activities involved in such operation and which eventually affect the productivity. The modeling of this uncertainty considers two variables:

1. The number of disruptions occurring in each of the activities in the operation,
2. The duration of the delays caused by each of these disruptions.

It is necessary to determine the statistical parameters describing the probability mass functions of both variables for each of the activities that the analyzed operation includes. Such parameters are meant to be the inputs to the construction simulation model formulated to analyze the operation. By this way, the effect of the PQM practices can be incorporated to the estimating of the productivity outputs obtained with this simulation model, since PQM practices are considered the root causes of disruptions, as explained before. Nevertheless, the data limitation may make the estimate of these statistical parameters unfeasible by means of probability-based input modeling techniques. Such techniques are thought to be unsuitable for modeling the previously described uncertainty because the existing data sets are usually insufficient for deciding on the corresponding probability distributions. Instead, the use of knowledge-based assessments elicited from experienced personnel represents a more suitable approach to model this uncertainty.

The implementation of the proposed modeling approach requires the development of a computer-based system that includes the following three subsystems.

A fuzzy expert system (FES) is to determine the quality levels of the project requirements, that is, the categories into which the nonconformance events are grouped. On the other hand, as seen in Figure 2, the corresponding performance levels of the PQM practices, that may be implemented in the organization or not, should be the inputs to this subsystem.

A fuzzy logic-based analytical approach (FLBAA) is to compute the statistical parameters (i.e., the mean and standard deviation) of both variables: the number of disruptions occurring in a given activity included in the operation and the durations of the delays caused by each of these disruptions. These statistical parameters should be obtained for every activity included in the construction operation under analysis. As seen in Figure 2, the quality levels of the project requirements obtained with the FES are meant to be the inputs to this subsystem.

The simulation project model of the specific operation is to be assessed, with which productivity estimates will be obtained. The abovementioned statistical parameters are needed as inputs to this simulation model.

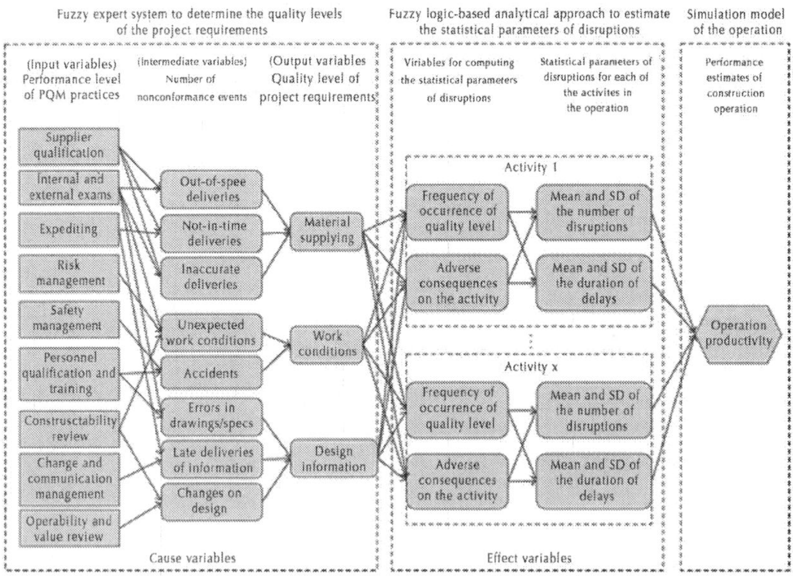


**Figure 2:** Simulation-based fuzzy logic approach to assessing the effect of PQM practices on the productivity of construction operations.

The two first subsystems are heuristic applications that facilitate the incorporation of knowledge-based assessments into the estimation of the statistical parameters required to run the simulation model of the construction operation. The integration of knowledge-based systems and simulation analysis has been studied since some decades ago [61–63], while the development of software engineering has made it easier to integrate these two modeling techniques so they can benefit each other. Such integrated approach has been successfully applied in different contexts: steel fabrication operations [64], chemical processes [65], service and manufacturing industries [9, 66], and emergency management systems [67]. Moreover, this approach has also been proven to be effective to solve construction problems, including applications to analyze construction operations [68–70], quantify the effect of accidents on construction operations [9], estimate accident risks on construction sites [8], estimate durations of construction activities [71], and estimate construction costs [5], among others. However, no evidences were found on the use of this approach to estimate the effect of quality management practices on the performance of construction operations.

The following sections provide details of each of the abovementioned subsystems but since the proposed modeling approach makes intensive use of fuzzy logic and fuzzy sets, a short introduction on such theory is first provided.

# FUZZY LOGIC AND FUZZY SETS

Fuzzy techniques have been widely used for solving a great diversity of construction problems [72]. Fuzzy logic uses fuzzy sets in order to facilitate the process of common sense reasoning with vague propositions dealing with linguistic variables, whose values are words or sentences expressed in natural or artificial languages [73]. For instance, the assessment of the number of nonconformities in a given construction activity may implement linguistic values such as very small, small, medium, large, or very large. Fuzzy sets use membership functions to represent the relationship between a range of possible values and a linguistic term. The degree with which each of these values is related to the linguistic term is defined within the interval zero (no membership) to one (full membership). By this way, it is possible to evaluate the "degree of truth" of propositions such as "x number of nonconformities is large." For instance, Figure 3 shows that the degree of membership of three nonconformities to the fuzzy set large is zero, while the degree of membership of five is one. In general, a fuzzy set A representing the degree of truth with which any element x (e.g., a given number of nonconformities) belongs to a set (e.g., a linguistic value) can be depicted by m discrete values $\mu A(x)$ as seen in

$$A = [x_1 / \mu_A(x_1), x_2 / \mu_A(x_2), x_3 / \mu_A(x_3), \ldots, x_m / \mu_A(x_m)], \qquad (2)$$

Where $\mu_A$ = (0 | 0.0, 1 | 0.0, 2 | 0.33, 3 | 0.66, 4 | 1.0, 5 | 0.5, 6 | 0.0, 7 | 0.0) $\mu_A(x)$ is the degree of membership [73].

For example, the fuzzy set medium in Figure 3 can be expressed as. Moreover, fuzzy numbers are simpler presentations of fuzzy sets that fulfill the properties of normality and convexity, which facilitate the fuzzy logic operations on fuzzy sets [73]. Triangular and trapezoidal membership functions, such as those shown in Figure 3, are common shapes for fuzzy numbers because little information is required to build them.

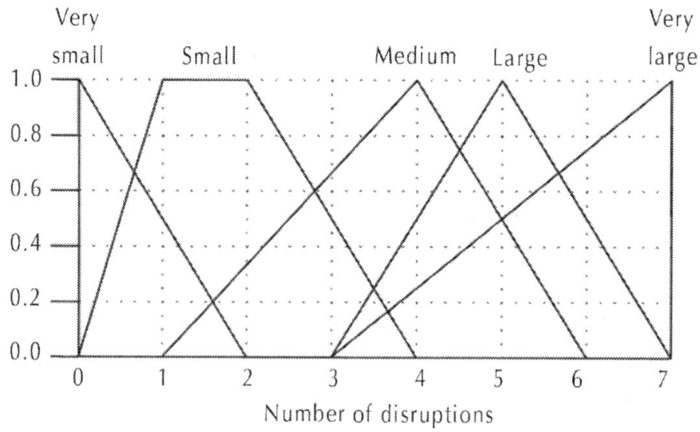


**Figure 3:** Membership functions for the number of disruptions in the pipe installation activity.

# FUZZY EXPERT SYSTEM TO DETERMINE THE QUALITY LEVEL OF PROJECT REQUIREMENTS

The purpose of the FES is to estimate, in fuzzy terms (i.e., very low, low, average, high, or very high), the quality levels of the project requirements (e.g., material supply, labor, and supervision). It is important to highlight that these estimates should be obtained at the project level and not at the operation level; that is, it is not necessary to obtain an estimate of the quality levels of requirements for every activity involved in the assessment. These estimates are needed as inputs to the FLBAA and, as proposed in this model (see Figure 2), they depend on the performance of the PQM practices that should be implemented to meet project quality. The model includes an intermediate variable, that is, the expected number of nonconformance events, which mediates the relationship between the project requirements and the PQM practices. The intention of such intermediate variable is to reduce the subjectivity of the assessments required for the estimating of the quality levels of project requirements. For example, the nonconformance events associated with the quality level of "material supply" may include "not in time deliveries," "inaccurate deliveries," and "out of specification deliveries," while the "supplier qualification" and the "Expediting" practices could prevent the occurrence of "not in time deliveries." Therefore, it is critical to identify the relationship between the PQM practices, the nonconformance events, and the project requirements that will be involved in the analysis. In this case, such identification can be done with root-cause analyses of the nonconformance events reported during past projects. First, the project requirements associated with the nonconformance events affecting a specific operation should be identified. Next, a Pareto analysis can be used to select the requirements with the highest frequency of nonconformance events; this would permit reducing the factors to be analyzed for the operation. Further, Pareto analyses can be carried out to select the most frequent types of nonconformance events associated with each of the previously selected requirements. Furthermore, a follow-up root-cause analysis will permit the identification of the PQM practices that should be implemented for avoiding the occurrence of the previously selected types of nonconformance events.

As seen in Figure 2, the variables integrated in the FES depict the cause variables in the proposed model and should be modeled as follows.

The performance levels of the PQM practices should be represented as input interfaces and can be expressed as either an estimated quality index (see (1)) or a linguistic term. The five linguistic terms proposed to assess this variable include no formal approach, reactive approach, stable and formal approach, continual improvement approach, and best-in-class performance. Therefore, the input interfaces should be built accordingly in order to use crisp values or fuzzy values as inputs to the system. Input interfaces that use crisp inputs require the representation of performance levels as membership functions, such as those shown in Figure 4.

The number of a given type of nonconformance event should be represented as an intermediate variable, which attains values during the inference process that are not needed as outputs of the model; their only function is passing information from one rule block to another. This information is conveyed during the inference process in the form of fuzzy (i.e., linguistic) values, which means that they do not need to be represented as membership functions. The linguistic terms used for assessing this variable are very low, low, average, high, and very high.

The quality levels of the project requirements should be modeled as output interfaces. Since the outputs to be obtained from the FES are required as fuzzy values, the inference process does not include the defuzzification of the fuzzy output. The linguistic terms used for assessing this variable are very poor, poor, average, good, and very good. Although the membership functions of these linguistic values are not required in the FES inference process, they should still be generated as they will be required by the inference process in the FLBAA. Figure 5 is a sample of these membership functions.

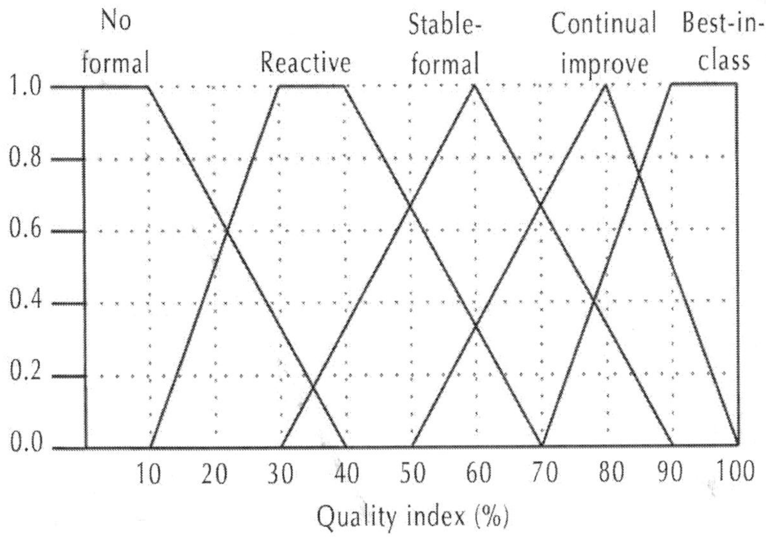


**Figure 4:** Membership functions representing the linguistic terms used to assess the performance level of the PQM practices.

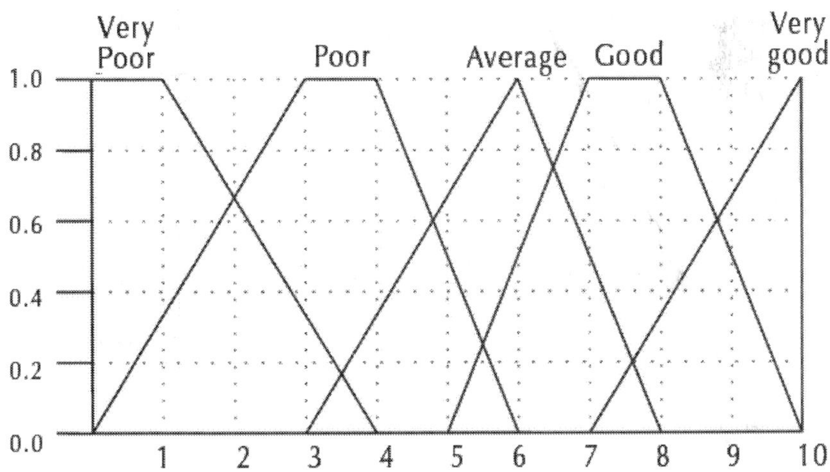


**Figure 5:** Membership functions for evaluating the quality level of project requirements.

Moreover, the inference process for determining the quality levels of project requirements requires two different sets of rule blocks that represent the following.

The effect of the performance levels of the PQM practices on the number of a given type of nonconformance event: for example, this includes rules like "IF the performance level of Personnel Qualification & Training is Reactive Approach AND the performance level of Safety Management is Stable Formal THEN the Number of Accidents is high."

The effect of the number of nonconformance events on the quality level of the corresponding project requirements: this includes rules like "IF the number of Accidents is high AND the number of Unexpected Work Conditions is Low THEN the quality level of Work Conditions is Poor."

Experienced project managers involved in the organization and the kind of project for which the analysis is being conducted, such as general supervisors and program managers, could provide the knowledge-based assessments required to produce the abovementioned membership functions and inference rules. The introduction to the inference process that is inherent to any FES can be found in G. Bojadziev and M. Bojadziev [73].

Moreover, the FES built with the above-explained approach should be validated. This should be done in order to further reduce the effect that the ambiguity of the assessments, provided by managers participating as developers of the proposed system, could have on the outcomes.

# FUZZY LOGIC-BASED ANALYSIS TO DETERMINE THE STATISTICAL PARAMETERS OF ACTIVITY DISRUPTIONS

The purpose of the FLBAA is determining, for each of the construction activities in the analyzed operation, the statistical parameters (i.e., the mean and standard deviation) of two variables: the number of disruptions occurring in a given activity included in the operation and the durations of the delays caused by each of these disruptions. These variables are associated with the disruptions occurring during the execution of the construction activities. Linguistic terms such as very low, low, average, high, and very high are used for assessing these two variables.

It is important to highlight that these estimates should be obtained at the operation level, which means that it is necessary to estimate the statistical parameters for every activity involved in the operation under assessment.

The proposed approach analyzes the combined effect of the quality levels of the project requirements on the performance (i.e., productivity in this case) of a given activity, for example, the combined effect of the good quality level of labor, the average quality level of supervision, and the poor quality level of material supply, on the productivity of the excavation activity.

The formulation of the FLBAA is based on the methodology proposed by Ayyub and Haldar [74], whose efficacy was proven in the solving of different construction problems such as the scheduling of projects [74], the assessment of safety in construction operations [75], the selection of construction strategies [76], the estimation of accident risk in construction projects [8], the analysis of delays in construction operations [77], and the quantification of the effects of accidents on construction operations [9].

This methodology permits obtaining the normally distributed probability mass function of the occurrence of a given event, based on the subjective assessments of the state of the factors affecting the event. In the proposed model, the project requirements depict such factors. By this way, the estimate of the required statistical parameters can be based on the assessment of the effect of the quality levels of project requirements on the performance of the activities involved in the analyzed operation.

The FLBAA brings into play the following variables:

1. Q**:** the quality levels of the project requirements involved in the operation under analysis,
2. F: the frequency of occurrence of the quality levels of the project requirements,

3. C: the level of adverse consequences due to the effect of the quality levels () on the activity under analysis,
4. N: the number of disruptions counted during the performance of the activity,
5. D: the duration of the delay due to the occurrence of disruptions in the activity.

As shown in Figure 2, F, C, N, D and represent the effect variables, as they depend on the performance of the construction project requirements and the PQM practices. The purpose of the FLBAA is to deliver the mean value and the standard deviation of both the number of disruptions counted during the performance of the activity (N) and the duration of the delay due to the occurrence of disruptions in the activity (D), for each of the activities involved in the operation under analysis. This involves the implementation of the following procedures.

1. Generate the membership functions of the abovementioned variables and the inference rules that will regulate the inference process. The development of these membership functions and inference rules should be specific to each of the activities within the analyzed operation and based on knowledge-based assessments provided by informed personnel directly participating in the undertaking of the activities, such as supervisors and laborers. Table 1 contains the linguistic terms commonly used in construction for assessing the effect variables. Every one of these terms should be represented as a membership function. For example, Figure 5 shows a sample of the membership functions representing the linguistic values of the quality levels (). In this case, psychometric scales with values ranging from zero to ten were thought to be appropriate domains for constructing the membership functions of Q, F, and C, while natural numbers were more appropriate for the membership functions representing the linguistic values of N (see, e.g., Figure 3) and D.

**Table 1**: Linguistic terms used for assessing the variables involved in the FL-BAA.

| Variable | Linguistic terms | | | | |
|---|---|---|---|---|---|
| Q | Very poor | Poor | Average | Good | Very good |
| F | Very unusual | Unusual | Often | Usual | Very usual |
| C | Very mild | Mild | Medium | Severe | Very severe |
| N | Very small | Small | Medium | Large | Very large |
| D | Very small | Small | Medium | Large | Very large |

Footnote

Note. : Q quality level of the project requirements, F frequency of occurrence, C adverse consequences, N number of disruptions, and: D duration of delays.

Moreover, Table 2 contains a sample of the rules that describe the effect of the different states (i.e., the quality levels) of the "work conditions" on the performance of the excavation activity. It is necessary to determine how frequently each of these states occurs during the performance of the activity, as well as the magnitude of the consequences on the activity due to such states. For example, the frequency of occurrence of the very poor quality level of "work conditions" in the excavation activity is unusual and the consequences on this activity are believed to be very severe, which eventually may generate a very large number of disruptions and large duration of delays in the activity. Table 2 also includes the corresponding linguistic values for F, C, N, and D when Q is poor, average, good, and very good.

**Table 2:** Sample of inference rules for analyzing the effect of work conditions on the excavation activity

| Q | F | C | N | D |
|---|---|---|---|---|
| Very poor | Unusual | Very severe | Very large | Large |
| Poor | Often | Severe | Large | Large |
| Average | Often | Medium | Medium | Medium |
| Good | Often | Mild | Small | Small |
| Very good | Unusual | Very mild | Small | Small |

Footnote

Note. : quality level of the project requirements, : frequency of occurrence, : adverse consequences, : number of disruptions, and : duration of delays.

Compute, given a combination of quality levels (Q) in the operation being analyzed, the following.

2. Compute the membership functions resulting from the fuzzy relations that combine the membership functions of the corresponding Q and C values of every one of the project requirements in the operation, as seen in

$$\mu_Q \times c(x_i, y_j) = \min\left[\mu_Q(x_i), \mu_C(y_j)\right], \tag{3}$$

Where $x_i$ and $y_j$ are values within psychometric scales from zero to ten used for assessing Q and ,C respectively, while $\mu_Q(x_i)$ and $\mu_c(y_j)$ are the respective degrees of membership.

a. Compute the membership function $T_1$ representing the total effect of the Q values on the performance of an activity, which results from the fuzzy union (U) of all the n sets of Cartesian relations obtained with (3), as seen in

$$T_1 = \bigcup_{i=1}^{n} \left[ \mu_Q \times c\left(x_i, y_j\right) \right], \tag{4}$$

Where n is the number of project requirements involved in the operation under analysis.

b. Compute the membership functions resulting from the fuzzy relations combining the corresponding F and N values of each of the project requirements involved in the operation, as seen in

$$\mu_F \times_N \left(f_j, r_k\right) = \min\left[\mu_F\left(f_j\right), \mu_N(r_k)\right], \tag{5}$$

where $f_j$ are values within a psychometric scales from zero to ten used for assessing F and $r_k$ are natural numbers used for appraising N (i.e., $r_k = 1, 2, 3, \ldots, n$), while $\mu_F\left(f_j\right)$ and $\mu_N\left(r_k\right)$ are the respective degrees of membership.

c. The membership function $T_2$ representing the total effect of the F values on the number of nonconformities results from the fuzzy union (U) of the n sets of Cartesian relations obtained with

$$T_2 = \bigcup_{i=1}^{n} \left[ \mu_F \times_N \left(f_j, r_k\right) \right], \tag{6}$$

Where n is the total number of project requirements involved in the operation.

d. The total effect of C on N in a given activity can be expressed with

$$R_N = (C_1 \times N_1) \cup \left(C_2 \times N_2\right) \ldots \cup \left(C_n \times N_n\right), \tag{7}$$

Where $C_n \times N_n$ are fuzzy relations based on fuzzy condition expressions represented by statements assessing the relation between each possible C value and a corresponding expected value.

For example, if $C_1$ is severe, then $N_1$ is large; if $C_2$ is medium, then $N_2$ is average, and so on. Such relations can be determined with experts' assessments of the performance of the given construction operation.

e. The membership function M resulting from the fuzzy composition relation between $T_1$ and can be calculated as seen in

$$M = \mu_{T_1 \cdot R_N}(x_i, r_k) = \max_{yj}\{\ \min\left[\mu_{T_1}(x_i, y_j), \mu_{R_N}(y_j, r_k)\right]\ \}, \tag{8}$$

Where $\mu_{T_1}.R_N(x_i, r_k)$ is the membership function of the fuzzy composition relation between $T_1$ and $R_N, \mu_{T_1}(x_i, y_j)$, is the membership function obtained with (4), and $\mu_{R_N}(y_j, r_k)$ is the membership function resulting from the fuzzy relation obtained with (7).

f. The membership function of the fuzzy joint effect of Q and F on the expected N can be developed as seen in

$$\mu_{M/T_2}(x_i, y_j)(r_k) = \min\left[\mu_M(x_i, r_k), \mu_{T_2}(f_j, r_k)\right], \tag{9}$$

Where $\mu_M(x_i, r_j)$ is the membership function of the effect of the quality levels of project requirements (Q) on the expected number of disruptions (N), which was obtained with (8), while $\mu_{T_2}(f_j, r_k)$ is the membership function of the effect of F on the expected N in the activity, obtained with (6). This fuzzy joint relationship will produce m number of matrices, each of which corresponds to an element $r_k$ in the subset of possible values of N. Moreover, the values of N may vary according to the uniqueness of the project and activity being analyzed. For example, Figure 3 shows that, for a given project, the subset of values $r_k$ ranges from zero to seven in the pipe installation activity.

3. Calculate the probability of occurrence of each element $r_k$ within the subset of possible values of N. In order to calculate this figure, Oliver's and Fayek [77] deduced from Ayyub and Haldar [74] the following:

$$P(N = rk) = \frac{\mu SN(rk)}{\sum_{k=1}^{m} \mu SN(rk)'} \tag{10}$$

Where N is the expected number of disruptions, $P(N = r_k)$ is the probability of occurrence of N being element $r_k, \mu_S(r_k)$ is the membership value of each element $r_k$ in the subset $S_N$ comprising the possible values of N, and m is the number of elements in the subset $S_N$ Further details for determining the membership value of $r_k$ can be found in Ayyub and Eldukair [75].

Subsequently, the mean value of the number of disruptions () and the corresponding standard deviation $(\sigma_N)$ can be, respectively, calculated with

$$\bar{N} = \sum_{k=1}^{m} (r_k) \times P(N = r_k)$$

$$\sigma_{[N]} = \sqrt{\left[ \sum_{k=1}^{m} (r_k)^2 \times P(N = r_k) \right] - (\bar{N})^2}. \tag{11}$$

This procedure should be carried out to compute the mean and standard deviation of N in each of the activities in the project. Moreover, it should also be used to estimate the statistics of D in such activities.

# KNOWLEDGE ELICITATION AND REPRESENTATION

This research adopted the procedures proposed by Fayek and Sun [78] to generate the membership functions of the variables involved in the case study model. This procedure included the following steps.

1. The linguistic terms commonly used by managers to assess the performance level of the variables involved in the model were determined, as seen in Table 3. Meetings with quality managers in the organization used for the sample application permitted

a significant amount of discussion that gave relevant advice regarding the assessment of the variables.

2. The domains or universes of discourse and their respective ranges of values were determined for each of the aforementioned variables, as seen in Table 3. In order to establish a general numerical guideline for the assessment of the variables, it was assumed that the extreme value of no formal approach, very poor, very mild, and very unusual was zero. On the other hand, it was assumed that the extreme value for best-in-class performance is 100% for assessing the performance level for PQM practices and very good, very severe, and very usual are ten for assessing the quality level of project requirements (Q), the adverse consequences (C), and the frequency of occurrence (F), respectively.
3. A questionnaire was developed in order to conduct the elicitation of the knowledge-based assessments needed for the development of the membership functions. By this way, the interviewees were asked to assign the appropriate numerical values for each of the linguistic variables in Table 3. For example, a sample question formulated to determine the numerical values for the linguistic terms assessing the performance levels of the PQM practices is illustrated in Table 4, while a sample question assessing the quality level of construction resources is shown in Table 5. The questionnaire was applied during a meeting with eight project managers of the construction organization that participated in the sample application.
4. The assessments of all respondents were accumulated for the development of the membership functions of the variables. The development of membership functions was based on the frequency with which each value, within the respective domain, was favored by the interviewees participating in the survey. A preliminary membership value was determined by the average degree of belief that the respondents had regarding the belonging of a value or range of values to each the linguistic terms used to evaluate the respective variables. This approach is illustrated in Table 6, which shows the ranges of values within the proposed subjective scale (from 0 to 100%) that each of the eight interviewees believed should correspond to the linguistic term "no formal approach" for the performance level of PQM practices. The

opinions favoring each value were summed and the result was standardized by dividing them by the total number of interviewees in order to obtain the preliminary membership function for the linguistic term, which can be expressed in its $x_i \mid \mu(X_i)$ form as $[0\% \mid 1.0, 10\% \mid 1.0, 20\% \mid 0.67, 30\% \mid 0.33, 40\% \mid 0.00]$. Likewise, all the opinions expressed for the other variables were developed into preliminary fuzzy membership functions.

5. The preliminary membership functions obtained with the previous procedure were adjusted in order to obtain triangular or trapezoidal shaped membership functions. Triangular and trapezoidal membership functions were preferred in this case, because of the clarity they can provide to the analysis. The rules proposed by Fayek and Sun [78] for generating the triangular and trapezoidal membership functions were used in this case. Figure 4 illustrates the membership functions developed to represent the performance level of the PQM practices, while Table 7 includes the numerical version of such membership functions.

**Table 3:** Linguistic terms and domains for the assessment of variables in the model

| Variables | Domain | Domain range | Linguistic terms for assessment of variables | | | | |
|---|---|---|---|---|---|---|---|
| Performance level of PQM practices | Quality index | 0–100% | No formal approach | Reactive approach | Stable-formal system | Continual improvement emphasized | Best-in-class performance |
| Number of nonconformance events | Fuzzy | NA | Very low | Low | Average | High | Very high |
| Quality level of the project requirements (Q) | Fuzzy | NA | Very poor | Poor | Average | Good | Very good |
| Frequency level of occurrence of the quality level (F) | Psychometric | 0–10 | Very unusual | Unusual | Often | Usual | Very usual |
| Adverse consequence resulting of the quality level (C) | Psychometric | 0–10 | Very mild | Mild | Medium | Severe | Very severe |
| Number of disruptions in a given construction activity (N) | Number of disruptions | 0–+∞ | Very small | Small | Medium | Large | Very large |
| Duration of delays due to disruptions (D) | Duration of delay (hours) | 0–+∞ | Very small | Small | Medium | Large | Very large |

**Table 4:** Sample question to assign numerical values to linguistic terms assessing the performance level of PQM practices

| Questions | Answers | |
|---|---|---|
| (1) For the performance level of the supplier qualification practice | | |
| (a) From zero to what value would you consider as no formal approach? | From zero | To |
| (b) What range of values as reactive approach? | From | To |
| (c) What range of values as stable formal system? | From | To |
| (d) What range of values as continual improvement? | From | To |
| (e) From what value to 100 as best-in-class performance? | From | To 100% |
| (f) Other level (specify please) | From | To |

**Table 5:** Sample question to assign numerical values to linguistic terms assessing the quality level of project requirements

| Questions | Answers | |
|---|---|---|
| (1) For the quality level of material supply | | |
| (a) From zero to what value would you consider it as very poor? | From zero | To |
| (b) For poor? | From | To |
| (c) For average? | From | To |
| (d) For good? | From | To |
| (e) From what value to 10 for very good? | From | To ten |

**Table 6:** Example of the calculations for the development of membership functions

| Opinions for no formal approach | Elements within the subjective scale (in %) | | | | | | | | | | |
|---|---|---|---|---|---|---|---|---|---|---|---|
| 0 | 10 | 20 | 30 | 40 | 50 | 60 | 70 | 80 | 90 | 100% | |
| Interviewee A | x | x | x | | | | | | | | |
| Interviewee B | x | x | x | | | | | | | | |
| Interviewee C | x | x | x | x | | | | | | | |
| Interviewee D | x | x | | | | | | | | | |
| Interviewee E | x | x | x | x | | | | | | | |
| Interviewee F | x | x | x | | | | | | | | |
| Interviewee G | x | x | | | | | | | | | |
| Interviewee H | x | x | x | x | | | | | | | |
| Sum of opinions | 8 | 8 | 6 | 3 | 0 | 0 | 0 | 0 | 0 | 0 | 0 |
| Standardized frequency | 1.0 | 1.0 | 0.75 | 0.38 | 0.00 | 0.00 | 0.0 | 0.0 | 0.0 | 0.0 | 0.0 |

**Table 7:** Sample of membership functions for the performance level of PQM practices.

| Linguistic terms | Fuzzy membership functions | | | | | | | | | | | |
|---|---|---|---|---|---|---|---|---|---|---|---|---|
| | 0% | 10% | 20% | 30% | 40% | 50% | 60% | 70% | 80% | 90% | 100% | Shape |
| No formal approach | 1.00 | 1.00 | 0.67 | 0.33 | 0.00 | 0.00 | 0.00 | 0.00 | 0.00 | 0.00 | 0.00 | Trap. |
| Reactive approach | 0.00 | 0.00 | 0.50 | 1.00 | 1.00 | 0.67 | 0.33 | 0.00 | 0.00 | 0.00 | 0.00 | Trap. |
| Stable formal system | 0.00 | 0.00 | 0.00 | 0.00 | 0.33 | 0.67 | 1.00 | 0.67 | 0.33 | 0.00 | 0.00 | Triang. |
| Continual improvement | 0.00 | 0.00 | 0.00 | 0.00 | 0.00 | 0.00 | 0.00 | 0.50 | 1.00 | 0.50 | 0.00 | Triang. |
| Best-in-class performance | 0.00 | 0.00 | 0.00 | 0.00 | 0.00 | 0.00 | 0.00 | 0.00 | 0.50 | 1.00 | 1.00 | Trap. |

Likewise, all the preliminary membership functions obtained with the raw data were

Adjusted and developed into triangular or trapezoidal shapes. This should be done for all variables listed in Table 3. It is important to highlight that the membership functions of the first three variables in this table should be generated based on the assessment of the conditions at the project level, while, for the membership function of the last four other variables, the operation level should be considered (i.e., it is necessary to generate different membership functions for each construction activity).

# SIMULATION MODEL FOR SAMPLE APPLICATION

The sample application of the proposed system analyzes the effect of the PQM practices implemented in the construction of a section of a new sanitary sewer system that includes the installation of 1320 feet (402.3 meters) of 24-inch concrete pipe pieces with a standard length of 3.0 meters. This entailed the undertaking of the open-cut construction operation that included the activities listed in Table 8. Since the advance rates of the activities have been assumed to be triangularly distributed, they were defined with the most optimistic, most pessimistic, and most likely estimates whose computation was based on the expertise of on-site managers.

**Table 8:** Durations and project requirements of the activities.

| Activity | Project requirements | Advance rate (min/m) | Distance buffer (m) |
|---|---|---|---|
| Excavation | 1 backhoe<br>1 survey crew | (2.9, 4.2, 6.1) | 20 |
| Detailing excavation | 6 laborers<br>1 survey crew | (3.1, 4.0, 5.2) | 40 |
| Bedding | 1 loader<br>2 types of vibratory equipment<br>6 laborers | (1.9, 3.8, 4.7) | 20 |
| Pipe installation | 1 loader<br>2 pipeman<br>1 laborer | (3.2, 4.5, 5.5) | 20 |
| Finishing of bedding | 1 loader<br>2 types of vibratory equipment<br>6 laborers | (2.5, 3.9, 5.2) | 40 |
| Backfilling | 1 dozer<br>1 type of compaction equipment | (2.7, 3.6, 4.8) | |

In order to simplify this sample application, the analysis only included the cause variables listed in Figure 2. A Pareto analysis of the historical data available in the organization permitted identifying the three most critical project requirements affecting the performance of open-cut construction operations, namely, the material supplying, the work conditions, and the design information. Besides, root-cause analyses allowed the identification of the corresponding associated nonconformities and PQM practices. It should be noted that the identification of the PQM practices brought into play the prevention and appraisal activities included in the Quality Performance Management System proposed by the Construction Industry Institute (CII), cited in Ledbetter [79], and the key principles and recommended practices articulated by the CII and the Business Roundtable's Construction Industry Cost Effectiveness project [80].

The implementation of the FES used Fuzzytech, a software development tool for fuzzy logic analysis, in order to facilitate the inferring of the quality levels of the project requirements (). This implementation assumed the input variables (i.e., the performance levels of the PQM practices) as mutually independent, as well as the intermediate variables (i.e., the number of nonconformance events). Therefore, the PROD and the MIN operators were used to aggregate the corresponding effect of the input and intermediate variables on the intermediate and output (i.e., the quality levels of project requirements) variables, respectively. Moreover, the MAX operator was applied for aggregating the results of both the intermediate and output variables. Further clarification of the fuzzy operators used for aggregating the inputs and outputs in the FES can be found in Tsoukalas and Uhrig [81]. On the other hand, the procedure on which the FLBAA is based was coded in MATLAB in order to facilitate the computing of the statistical parameters of the two variables associated with the disruptions: the number of disruptions occurring in each of the activities in the operation and the duration of the delays caused by each of these disruptions. As explained before, the implementation of the FES and the FLBAA required the participation of experienced project managers and on-site employees, respectively, for the construction of the corresponding membership functions and inference rules. In addition, a discrete-event simulation model of the open-cut construction process was built up with Simphony. NET, a simulation platform developed by the NSERC/Alberta Construction Industry Research Chair at the University of Alberta

## Sensitivity Analysis

The initial part of this sample application used the FES for analyzing the sensibility of the quality levels of project requirements to the performance levels of the PQM practices. The intention of this analysis is identifying the PQM practices that should be favored for improvement, given the significance of their effect on the quality levels. The first treatment included setting at 0% the quality index (see (1)) of every other of the PQM practices that affect a given resource while setting at 100% the quality index of the other PQM practices. This implied that the lower the resulting quality level is, the more significant the effect of the PQM practice set at 0% on the corresponding project requirement is. The second treatment was conducted in order to reinforce the conclusions obtained with the first one. This time, the quality index of one of the PQM practices was set at 100%, while the other practices were set at 0%. In this case, the higher the resulting quality level is, the more significant the effect of the PQM practice set at 100% is. This also permitted identifying the practices that would have the most significant contribution towards the performance improvement of each of the project requirements. Based on the findings obtained, the significance of the effect of the PQM practices on the quality levels of the project requirements was ranked as low, medium, or high, as shown in Table 9. For example, this suggests that the effect of the supplier qualification practice on the quality level of material supply is high, while the effect of the internal and external examinations and the expediting practices is medium. A more efficient improvement process of the PQM system should prioritize increasing the performance of those PQM practices with the highest rankings of significance.

**Table 9:** Significance of the effect of the PQM practices on the project requirements

| Project requirements | PQM practices | Significance on project requirements |
|---|---|---|
| Material supply | Expediting | Medium |
| Supplier qualification | High | |
| Internal and external examinations | Medium | |

| | | |
|---|---|---|
| Design information | Change and communication | Medium |
| Constructability review | Medium | |
| Internal and external examinations | Medium | |
| Expediting | Low | |
| Operability and value review | Low | |
| Personnel qualification and training | High | |
| Work conditions | Constructability review | Low |
| Internal and external examinations | Medium | |
| Personnel qualification and training | High | |
| Risk management | Low | |
| Safety management | High | |

## Simulation Outputs

The three hypothetical scenarios in Table 10 were formulated in order to illustrate the estimate of the effect of improving the performance levels of the PQM practices. This table contains four sections, each of which describes one of the different types of data processed within the computer-based system. The first section contains the inputs (i.e., the performance levels of the PQM practices) to the FES, which in this case have been formulated using fuzzy values (i.e., linguistic terms). In order to facilitate the analysis of the outputs, each of these fuzzy values features an equivalent crisp value, which is the quantitative depiction (displayed in parentheses) of the linguistic term used to assess the performance level of a given PQM practice. Though they are similar to the estimates of the quality indexes, the equivalent crisp values should be obtained from the defuzzification of the membership functions representing the linguistic terms instead of the self-assessment approach explained before. Such defuzzification can be based on the method proposed by Mason and Kahn [82]. For example, as obtained from the defuzzification of the membership functions in Figure 5, Table 10 shows that the equivalent crisp value of no formal approach is 21%, reactive approach is 38%, stable formal system approach is 60%, continual improvement approach is 76%, and best-in-class performance is 87%. On the other hand, the second section contains the analyzed requirements' quality levels that were inferred with the sample FES, based on the input settings of the different scenarios. These outputs were fed to the FLBAA in order to compute, for each of the analyzed activities, the statistical parameters shown in the third section of the table. Finally, the fourth section contains the mean productivity estimates obtained as outputs of the simulation project model formulated for analyzing the sample operation. In this case, fifty simulation runs were completed for computing the mean productivity estimate in each one of the scenarios.

**Table 10:** Settings of the scenarios and their corresponding simulation outputs

| Scenarios → | | 1 | 2 | 3 |
|---|---|---|---|---|
| Input settings (Performance Levels of the PQM practices) | | | | |
| Expediting | | Stable (60%) | Stable (60%) | Stable (60%) |
| Supplier qualification | | Reactive (38%) | Stable (60%) | Stable (60%) |
| Int. and ext. examinations | | Reactive (38%) | Stable (60%) | Stable (60%) |
| Change and communications | | Reactive (38%) | Stable (60%) | Stable (60%) |
| Operability and value review | | Reactive (38%) | Stable (60%) | Reactive (38%) |
| Constructability review | | Reactive (38%) | Stable (60%) | Stable (60%) |
| Personnel qualification | | Stable (60%) | Stable (60%) | Continual (76%) |
| Risk management | | Reactive (38%) | Stable (60%) | Reactive (38%) |
| Safety management | | Stable (60%) | Stable (60%) | Continual (76%) |
| Quality levels of the project requirements (Q) | | | | |
| Material supply | | Poor | Good | Good |
| Design information | | Average | Good | Very good |
| Work conditions | | Good | Very good | Very good |
| Statistical parameters of the number of disruptions (N) | | | | |
| Excavation | $\overline{N}$ | 2.47 | 1.93 | 1.42 |
| | $\sigma_N$ | 1.11 | 0.81 | 1.07 |

| | | | | |
|---|---|---|---|---|
| Bedding | $\bar{N}$ | 2.22 | 1.45 | 0.95 |
| | $\sigma_N$ | 1.11 | 0.50 | 0.78 |
| Pipe installation | $\bar{N}$ | 2.72 | 1.94 | 1.44 |
| | $\sigma_N$ | 1.35 | 0.80 | 1.06 |
| Backfilling | $\bar{N}$ | 2.22 | 1.46 | 0.94 |
| | $\sigma_N$ | 1.11 | 0.50 | 0.78 |
| Statistical parameters of the duration of delays (D) | | | | |
| Excavation | $\bar{D}$ | 168.84 | 45.57 | 21.43 |
| | $\sigma_D$ | 85.58 | 46.00 | 28.75 |
| Bedding | $\bar{D}$ | 251.19 | 176.10 | 161.68 |
| | $\sigma_D$ | 114.00 | 84.99 | 87.56 |
| Pipe installation | $\bar{D}$ | 164.90 | 51.89 | 41.63 |
| | $\sigma_D$ | 85.95 | 48.65 | 47.27 |
| Backfilling | $\bar{D}$ | 164.90 | 87.69 | 56.64 |
| | $\sigma_D$ | 85.95 | 29.91 | 46.97 |
| Outputs and output analysis | | | | |
| MPE (meter/hour) | | 2.53 | 3.47 | 3.70 |
| Productivity increase | | — | 37.15% | 46.24% |
| TIPL | | — | 132% | 120% |
| RPI | | — | 0.28 | 0.39 |

Footnote

MPE: mean productivity estimates; TIPL: total increase of the performance levels; RPI: relative productivity improvement.

MPE: mean productivity estimates; TIPL: total increase of the performance levels; RPI: relative productivity improved

## Output Analysis

The analysis of the outputs assumes that, in order to improve the operation's current performance, it is necessary to increase the performance levels of the PQM practices. Therefore, it is feasible to compute the total increase of the performance levels, which results in the sum of the variations between the equivalent crisp values that depict the improved and the current performance levels. For example, assume that scenario 1 in Table 10 features the current states of the PQM practices; hence, the total increase of the performance levels in scenario 2 is 132% given that the equivalent crisp values increased by 22% (i.e., the variation between 60% in scenario 2 and 38% in scenario 1) in six of the PQM practices (i.e., 22% × 6). Moreover, the output analysis also includes the computing of the productivity increase that is expected from the increase of the performance levels of the PQM practices. The productivity increase in a given scenario is the ratio of the mean productivity estimates obtained in that scenario to the mean productivity estimates obtained in the corresponding previous scenario. For example, the productivity increase in scenario 2 is 37.15% given that the mean productivity estimates is 3.47 m/hr in scenario 2 and 2.53 m/hr in scenario 1.

The purpose of computing the productivity increase and the total increase of the performance levels is estimating the relative productivity improvement achieved with the input settings of a given scenario. The relative productivity improvement is the ratio of the productivity increase to the effort implemented in the improvement of the PQM system (i.e., the total increase of the performance levels), and it serves to benchmark the outputs obtained in two or more alternative scenarios. This way, the relative productivity improvement may facilitate the recognition of the most efficient strategies for improving a given PQM system. For example, suppose that while the PMLs listed as input settings in scenario 1 correspond to the current state of the PQM practices, the ones in scenarios 3 and 4 depict potential targets for improving the performance currently attained. Therefore, in order to decide which of the two scenarios represents the best alternative, the outputs obtained in scenarios 3 and 4 have been evaluated with reference to those obtained in scenario 2. On one hand, scenario 3 features a conventional strategy in which all the PQM practices have been set at the stable and formal approach performance level. On the other hand, scenario 4 represents a more conscious strategy in which the upgrading of the practices with a high or medium significance on the quality level of the project requirements (refer to Table 9) has been prioritized. For example, while key PQM practices such as personnel qualification and training and safety management have been upgraded at the continual improvement approach performance level, less relevant practices such as the operability and value review and the risk M management have been kept at the reactive approach level. As seen in Table 10, the input settings in scenario 3 have produced higher quality levels of project requirements and, hence, better statistical parameters than those obtained with the inputs in scenario 2. This finally resulted in scenario 3 featuring a higher mean productivity estimate (i.e., 3.70 m/hr) than that obtained in scenario 2 (i.e., 3.47 m/hr). Moreover, it should be noted that the total increase of the performance levels in scenario 3 (i.e., 120%) is lower than that in scenario 2 (i.e., 132%). Therefore, the relative productivity improvement resulted significantly higher in scenario 3 (i.e., 0.39 against 0.28 in scenario 2), which means that it represents a more efficient alternative to improve the PQM system.

# LIMITATIONS OF THE PROPOSED APPROACH AND FUTURE WORK

It is important to mention that since the purpose is the proposed modeling approach is assessing exclusively the effect of the PQM factors on the performance of the construction operations, this modeling approach excludes the effect of other factors affecting the construction operations. Only factors related to project quality management were considered in this modeling approach and the obtained estimates should not be considered accurate as to take decisions regarding the scheduling of the project. They should be used to support decision making concerning quality performance improvement. However, future research could focus on the incorporation of this approach to a more comprehensive model that integrates the effect of a full list of factors on the estimating of construction performance. This would permit obtaining more accurate estimates that could be used for project scheduling purposes.

The outcome of this study demonstrated the comprehensiveness of the proposed modeling approach to integrate the factors associated with PQM; however, this can also lead to further work in this research area. The outcomes derived in future could benefit practice, and future research work should contemplate the inclusion of the costs associated with the upgrading of each of the PQM practices in the analysis. This would permit accomplishing an even more thoughtful decision-making process on the improvement of construction quality performance. Additional future research could focus on exploring that the input model defined by different experts may vary significantly to yield different results and the model by means of knowledge-based assessments for statistical parameters needs to be verified by actual site data in retrospective.

# SUMMARY AND CONCLUSIONS

The outcome of the research reported in this paper includes a unique methodology that integrates simulation modeling techniques with knowledge-based systems in order to assess the effect of project quality management on the performance of construction operations. The intention is to support decision making concerning the improvement of PQM systems and the increase of project performance.

The proposed simulation-based approach is based on the analysis of the factors involved in the relationship between the PQM practices and the performance of construction activities: the performance levels of the PQM practices to determine the quality levels of the project requirements, and these quality levels will determine the amount of disruptions that eventually will affect the productivity of construction operations. Moreover, this approach provides simulation estimates of productivity in which the effect of the performance of PQM is incorporated. Incorporating this effect in the productivity estimates entails computing, for each of the activities involved in the operation, the parameters that describe the statistical distributions of two variables: the number of disruptions that may occur during the given activity and the duration of the delays due to such disruptions. These statistical parameters are the inputs to the simulation model from which the productivity estimates of the operation will be obtained. In this case, the estimate of these statistical parameters involves two fuzzy logic-based applications: a FES for inferring the quality levels of the project requirements involved in the analyzed construction operation and a FLBAA that uses the outputs of the FES for computing the required statistical parameters. The use of fuzzy logic makes possible the incorporation of knowledge-based assessments into the estimate of the statistical parameters, which permits overcoming the limited availability of information that makes probability-based modeling techniques unsuitable for this purpose.

A sample application illustrated the assessment of the effect of a number of PQM practices on the productivity of an open-cut construction operation and demonstrated the predictive and explanatory capabilities of the proposed modeling approach. This included demonstrating that its capacity for exploring PQM improvement strategies can provide information on the mechanisms, interactions, and the most effective ways of achieving improved performance of the construction operations. Therefore, this clarified the usefulness of the obtained simulation productivity estimates in the understanding of the effect of PQM on the construction performance, as well as in the decision making concerned with the improvement of quality performance. The availability of this methodology also offers new approaches to the study of quality, supports the exploration of fundamental questions in construction project performance, and facilitates the identification of improvement opportunities.

## CONFLICT OF INTERESTS

The authors declare that there is no conflict of interests regarding the publication of this paper.

## ACKNOWLEDGMENTS

The research reported in this paper was supported by The National Council for Science and Technology (CONACYT—Mexico) and the Natural Sciences and Engineering Research Council/Construction Industry Research Chair (Canada). The authors would like to thank these institutions along with the companies that participated in this research.

# REFERENCES


1. H. Zhang, H. Li, and M. Lu, "Modeling time-constraints in construction operations through simulation," Journal of Construction Engineering and Management, vol. 134, no. 7, pp. 545–554, 2008.
2. N. N. Chehayeb and S. M. AbouRizk "Simulation-based scheduling with continuous activity relationships," Journal of Construction Engineering and Management, vol. 124, no. 2, pp. 107–115, 1998.
3. D.-E. Lee and D. Arditi, "Automated statistical analysis in stochastic project scheduling simulation," Journal of Construction Engineering and Management, vol. 132, no. 3, pp. 268–277, 2006.
4. Y. Mohamed, D. Borrego, L. Francisco, M. Al-Hussein, S. AbouRizk, and U. Hermann, "Simulation-based scheduling of module assembly yards: case study," Engineering, Construction and Architectural Management, vol. 14, no. 3, pp. 293–311, 2007.
5. A. A. Shaheen, A. R. Fayek, and S. M. AbouRizk "Fuzzy numbers in cost range estimating,"Journal of Construction Engineering and Management, vol. 133, no. 4, pp. 325–334, 2007.
6. M. Lu, S.-C. Lau, and C.-S. Poon, "Simulation approach to evaluating cost efficiency of selective demolition practices: case of Hong Kong's Kai Tak Airport demolition," Journal of Construction Engineering and Management, vol. 135, no. 6, pp. 448–457, 2009.
7. S. Fernando, J. Tan, K. C. Er, and S. AbouRizk "Simulation: a tool for effective risk management," in Proceedings of the 5th Construction Specialty Conference of the Canadian Society for Civil Engineering, Moncton, Canada, June 2003.
8. S. Lee and D. W. Halpin, "Predictive tool for estimating accident risk," Journal of Construction Engineering and Management, vol. 129, no. 4, pp. 431–436, 2003.
9. S. Lee, D. W. Halpin, and H. Chang, "Quantifying effects of accidents by fuzzy-logic- and simulation-based

analysis," Canadian Journal of Civil Engineering, vol. 33, no. 3, pp. 219–226, 2006.

10. A. Agbulos, Y. Mohamed, M. Al-Hussein, S. Abourizk, and J. Roesch, "Application of lean concepts and simulation analysis to improve efficiency of drainage operations maintenance crews," Journal of Construction Engineering and Management, vol. 132, no. 3, pp. 291–299, 2006.
11. R. Sacks, A. Esquenazi, and M. Goldin, "LEAPCON: simulation of lean construction of high-rise apartment buildings," Journal of Construction Engineering and Management, vol. 133, no. 7, pp. 529–539, 2007.
12. X. Mao and X. Zhang, "Construction process reengineering by integrating lean principles and computer simulation techniques," Journal of Construction Engineering and Management, vol. 134, no. 5, pp. 371–381, 2008.
13. P. Wang, Y. Mohamed, S. M. AbouRizk and A. R. T. Rawa, "Flow production of pipe spool fabrication: simulation to support implementation of lean technique," Journal of Construction Engineering and Management, vol. 135, no. 10, pp. 1027–1038, 2009.
14. D. J. Fisher and J. T. O'Connor, "Constructability for piping automation: field operations,"Journal of Construction Engineering and Management, vol. 117, no. 3, pp. 468–485, 1991.
15. A. Sobotka, "Simulation modelling for logistics re-engineering in the construction company,"Construction Management and Economics, vol. 18, no. 2, pp. 183–195, 2000.
16. S. T. Ng, J. Shi, and Y. Fang, "Enhancing the logistics of construction materials through activity-based simulation approach," Engineering, Construction and Architectural Management, vol. 16, no. 3, pp. 224–237, 2009.
17. C. Vidalakis, J. E. Tookey, and J. Sommerville, "Logistics simulation modelling across construction supply chains," Construction Innovation, vol. 11, no. 2, pp. 212–228, 2011.
18. W. Lo and M.-R. Yan, "Evaluating qualification-based selection system: a simulation approach,"Journal of Construction Engineering and Management, vol. 135, no. 6, pp. 458–465, 2009.

19. T. M. Zayed and D. W. Halpin, "Quantitative assessment for piles productivity factors," Journal of Construction Engineering and Management, vol. 130, no. 3, pp. 405–414, 2004.
20. T. M. Zayed and D. W. Halpin, "Simulation as a tool for pile productivity assessment," Journal of Construction Engineering and Management, vol. 130, no. 3, pp. 394–404, 2004.
21. L. Song and S. M. AbouRizk "Measuring and modeling labor productivity using historical data,"Journal of Construction Engineering and Management, vol. 134, no. 10, pp. 786–794, 2008.
22. T. M. Zayed and D. W. Halpin, "Pile construction productivity assessment," Journal of Construction Engineering and Management, vol. 131, no. 6, pp. 705–714, 2005
23. L. Song and S. M. AbouRizk "Virtual shop model for experimental planning of steel fabrication projects," Journal of Computing in Civil Engineering, vol. 20, no. 5, pp. 308–316, 2006.
24. E.-B. Lee and C. W. Ibbs, "Computer simulation model: construction analysis for pavement rehabilitation strategies," Journal of Construction Engineering and Management, vol. 131, no. 4, pp. 449–458, 2005.
25. M. M. Hassan and S. Gruber, "Simulation of concrete paving operations on interstate-74,"Journal of Construction Engineering and Management, vol. 134, no. 1, pp. 2–9, 2008.
26. J. Shi and S. AbouRizk "Continuous and combined event-process models for simulating pipeline construction," Construction Management and Economics, vol. 16, no. 4, pp. 489–498, 1998
27. S. M. AbouRizk and R. J. Wales, "Combined discrete-event/ continuous simulation for project planning," Journal of Construction Engineering and Management, vol. 123, no. 1, pp. 11–20, 1997.
28. C. Zhang, T. Zayed, and A. Hammad, "Resource management of bridge deck rehabilitation: Jacques cartier bridge case study," Journal of Construction Engineering and Management, vol. 134, no. 5, pp. 311–319, 2008.
29. T. M. Zayed and D. Halpin, "Simulation of concrete batch plant production," Journal of Construction Engineering and Management, vol. 127, no. 2, pp. 132–141, 2001.

30. H.-S. Lee, J.-W. Shin, M. Park, and H.-G. Ryu, "Probabilistic duration estimation model for high-rise structural work," Journal of Construction Engineering and Management, vol. 135, no. 12, pp. 1289–1298, 2009.

31. T. Zayed, M. R. Sharifi, S. Baciu, and M. Amer, "Slip-form application to concrete structures,"Journal of Construction Engineering and Management, vol. 134, no. 3, pp. 157–168, 2008.

32. Y. Ebrahimy, S. M. AbouRizk S. Fernando, and Y. Mohamed, "Simulation modeling and sensitivity analysis of a tunneling construction project's supply chain," Engineering, Construction and Architectural Management, vol. 18, no. 5, pp. 462–480, 2011

33. M. Al-Bataineh, S. AbouRizk and H. Parkis, "Using simulation to plan tunnel construction,"Journal of Construction Engineering & Management, vol. 139, no. 5, pp. 564–571, 2013.

34. A. Shahin, S. M. AbouRizk Y. Mohamed, and S. Fernando, "Simulation modeling of weather-sensitive tunnelling construction activities subject to cold weather," Canadian Journal of Civil Engineering, vol. 41, no. 1, pp. 48–55, 2014.

35. A. P. C. Chan and C. M. Tam, "Factors affecting the quality of building projects in Hong Kong,"International Journal of Quality & Reliability Management, vol. 17, no. 4-5, pp. 423–441, 2000.

36. K. N. Jha and K. C. Iyer, "Critical factors affecting quality performance in construction projects,"Total Quality Management and Business Excellence, vol. 17, no. 9, pp. 1155–1170, 2006.

37. E. W. M. Lam, A. P. C. Chan, and D. W. M. Chan, "Determinants of successful design-build projects," Journal of Construction Engineering and Management, vol. 134, no. 5, pp. 333–341, 2008.

38. A. Aghaie and K. Popplewell, "Simulation for TQM—the unused tool?" TQM Magazine, vol. 9, no. 2, pp. 111–116, 1997

39. R. Sonmez and J. E. Rowings, "Construction labor productivity modeling with neural networks,"Journal of Construction Engineering and Management, vol. 124, no. 6, pp. 498–504, 1998.

40. P. Love and D. J. Edwards, "Forensic project management: the underlying causes of rework in construction projects," Civil Engineering and Environmental Systems, vol. 21, no. 3, pp. 207–228, 2004.

41. C. Kontoghiorghes, "Examining the association between quality and productivity performance in a service organization," The Quality Management Journal, vol. 10, no. 1, pp. 32–42, 2003.

42. G. San, T.-C. Huang, and L.-H. Huang, "Does labour quality matter on productivity growth? the case of the Taiwanese manufacturing industry," Total Quality Management & Business Excellence, vol. 19, no. 10, pp. 1043–1053, 2008.

43. D. A. Langford, H. El-Tigani, and M. Marosszeky, "Does quality assurance deliver higher productivity?" Construction Management and Economics, vol. 18, no. 7, pp. 775–782, 2000.

44. W. E. Deming, Out of the Crisis, MIT Center for Advanced Engineering Study, Massachusetts Institute of Technology, Cambridge, Mass, USA, 1986.

45. R. D. Moen, T. W. Nolan, and L. P. Provost, Improving Quality through Planned Experimentation, McGraw-Hill Series in Industrial Engineering & Management, McGraw-Hill, New York, NY, USA, 1991.

46. F. Yasamis, D. Arditi, and J. Mohammadi, "Assessing contractor quality performance,"Construction Management and Economics, vol. 20, no. 3, pp. 211–223, 2002.

47. B. Sharma and D. Gadenne, "An inter-industry comparison of quality management practices and performance," Managing Service Quality, vol. 12, no. 6, pp. 394–404, 2002.

48. L. Lakhal, F. Pasin, and M. Limam, "Quality management practices and their impact on performance," International Journal of Quality & Reliability Management, vol. 23, no. 6, pp. 625–646, 2006.

49. Q. Su, Z. Li, S.-X. Zhang, Y.-Y. Liu, and J.-X. Dang, "The impacts of quality management practices on business performance: an empirical investigation from China," International Journal of Quality and Reliability Management, vol. 25, no. 8, pp. 809–823, 2008.

50. V. Kumar, F. Choisne, D. de Grosbois, and U. Kumar, "Impact of TQM on company's performance," International Journal of Quality & Reliability Management, vol. 26, no. 1, pp. 23–37, 2009.

51. X. Zu, "Infrastructure and core quality management practices: how do they affect quality?"International Journal of Quality & Reliability Management, vol. 26, no. 2, pp. 129–149, 2009.

52. P.-E. Josephson and Y. Hammarlund, "Causes and costs of defects in construction a study of seven building projects," Automation in construction, vol. 8, no. 6, pp. 681–687, 1999

53. M. G. Battikha, "Reasoning mechanism for construction nonconformance root-cause analysis,"Journal of Construction Engineering and Management, vol. 134, no. 4, pp. 280–288, 2008.

54. T. H. Willis and W. D. Willis, "A quality performance management system for industrial construction engineering projects," International Journal of Quality & Reliability Management, vol. 13, no. 9, pp. 38–48, 1996.

55. R. A. Orwig and L. L. Brennan, "An integrated view of project and quality management for project-based organizations," International Journal of Quality & Reliability Management, vol. 17, no. 4-5, pp. 351–363, 2000.

56. W. Carl and Y. C. Hurtado, "Quality management practice in capital projects," Transactions of the American Association of Cost Engineers, ABI/INFORM Global, pp. Q.4.1–Q.4.8, 1990.

57. T. A. Neese and W. B. Ledbetter, "Quality performance management in engineering/construction," Transactions of the American Association of Cost Engineers, ABI/INFORM Global, pp. A.2.1–A.2.10, 1991.

58. M. Barad and T. Raz, "Contribution of quality management tools and practices to project management performance," International Journal of Quality & Reliability Management, vol. 17, no. 4-5, pp. 571–583, 2000.

59. C. C. P. Kumar and C. Wolf, "Assessing project quality," AACE Engineers Transactions, vol. 2, pp. I.5.1–I.5.11, 1992.

60. S. M. AbouRizk and D. W. Harpin, "Probabilistic simulation studies for repetitive construction processes," Journal of

Construction Engineering and Management, vol. 116, no. 4, pp. 575–594, 1990

61. P. A. Fishwick and R. B. Modjeski, Knowledge-Based Simulation: Methodology and Application, vol. 4 of Advances in Simulation, Springer, New York, NY, USA, 1991.
62. P. A. Fishwick and P. A. Luker, Qualitative Simulation Modeling and Analysis, vol. 4 of Advances in Simulation, Springer, New York, NY, USA, 1991.
63. P. A. Fishwick, "An integrated approach to system modeling using a synthesis of artificial intelligence, software engineering, and simulation methodologies," ACM Transactions on Modeling and Computing Simulation, vol. 2, no. 4, pp. 307–330, 1992.
64. D. Stirling and S. Sevinc, "Combined simulation and knowledge-based control of a stainless steel rolling mill," Expert Systems with Applications, vol. 3, no. 3, pp. 353–366, 1991.
65. M. Eisenberg, "The kineticist's workbench: qualitative/quantitative simulation of chemical reaction mechanisms," Expert Systems With Applications, vol. 3, no. 3, pp. 367–377, 1991.
66. Y.-C. Chang and C.-C. Chen, "Knowledge-based simulation of bunkering services in the port of Kaohsiung," Civil Engineering and Environmental Science, vol. 23, no. 1, pp. 21–34, 2006.
67. J. Z. Hernández and J. M. Serrano, "Knowledge-based models for emergency management systems," Expert Systems with Applications, vol. 20, no. 2, pp. 173–186, 2001.
68. H. Zhang, C. M. Tam, and J. J. Shi, "Application of fuzzy logic to simulation for construction operations," Journal of Computing in Civil Engineering, vol. 17, no. 1, pp. 38–45, 2003.
69. A. Shaheen, A. Fayek, and S. AbouRizk "A framework for integrating fuzzy expert systems and discrete event simulation," in Proceedings of the ASCE Construction Research Congress, pp. 1–11, 2005.
70. Y. Mohamed and S. M. AbouRizk "Framework for building intelligent simulation models of construction operations," Journal of Computing in Civil Engineering, vol. 19, no. 3, pp. 277–291, 2005.
71. H. Zhang, H. Li, and C. M. Tam, "Fuzzy discrete-event simulation for modeling uncertain activity duration," Engineering,

Construction and Architectural Management, vol. 11, no. 6, pp. 426–437, 2004.

72. A. P. C. Chan, D. W. M. Chan, and J. F. Y. Yeung, "Overview of the application of "fuzzy techniques" in construction management research," Journal of Construction Engineering & Management, vol. 135, no. 11, pp. 1241–1252, 2009.
73. G. Bojadziev and M. Bojadziev, Fuzzy Logic or Business, Finance, and Management, vol. 12 ofAdvances in Fuzzy Systems: Applications and Theory, World Scientific, Singapore, 1997.
74. B. M. Ayyub and A. Haldar, "Project scheduling using fuzzy set concepts," Journal of Construction Engineering and Management, vol. 110, no. 2, pp. 189–204, 1984.
75. B. M. Ayyub and Z. A. Eldukair, "Safety assessment methodology for construction operations," inProceedings of the 5th International Conference on Structural Safety and Reliability (ICOSSAR '89), pp. 771–777, ASCE, New York, NY, USA, August 1989
76. B. M. Ayyub and A. Haldar, "Decisions in construction operations," Journal of Construction Engineering and Management, vol. 111, no. 4, pp. 343–357, 1985.
77. A. V. O. Oliveros and A. R. Fayek, "Fuzzy logic approach for activity delay analysis and schedule updating," Journal of Construction Engineering and Management, vol. 131, no. 1, pp. 42–51, 2005.
78. A. R. Fayek and Z. Sun, "A fuzzy expert system for design performance prediction and evaluation," Canadian Journal of Civil Engineering, vol. 28, no. 1, pp. 1–25, 2001.
79. W. B. Ledbetter, "Quality performance on successful project," Journal of Construction Engineering and Management, vol. 120, no. 1, pp. 34–46, 1994.
80. S. D. Anderson, "Project quality and project managers," International Journal of Project Management, vol. 10, no. 3, pp. 138–144, 1992, Reprinted from PMI '91 Seminar/ Symposium Proceedings, 28 September–2 October 1991, Dallas, TX, USA.
81. L. H. Tsoukalas and R. E. Uhrig, Fuzzy and Neural Approaches in Engineering, John Wiley & Sons, New York, NY, USA, 1997.

82. A. K. Mason and D. J. Kahn, "Estimating costs with fuzzy logic," AACE International Transactions, pp. 122–127, 1997.

Chapter 10

# Architect Critical Challenges As a Project Manager in Construction Projects: A Case Study

Mohammadreza Yadollahi[1], Mohammad Mirghasemi[2], Rosli Mohamad Zin[3], and Bachan Singh[1]

[1]Faculty of Civil Engineering, Universiti Teknologi Malaysia, 81310 Skudai, Johor, Malaysia

[2]Faculty of Built Environment, Universiti Teknologi Malaysia, 81310 Johor, Malaysia

[3]Construction Research Centre, Universiti Teknologi Malaysia, 81310 Johor, Malaysia

## ABSTRACT

All construction professionals such as civil, mechanical, and electrical engineers, quantity surveyors, and architects have important roles in the construction process. Among these, architects are frequently appointed as a project manager (PM). The role of a PM will drive the success of the projects implementation. Therefore, the capability of an architect as a PM (ArPM) is critical in reducing challenges encountered. Accordingly, the identification of these challenges is an important task in selecting an appropriate ArPM. The aim of this study is to identify the most critical challenges faced by an ArPM for construction projects. The data were collected through questionnaires and interviews with architects and professionals in the Malaysian construction industry. Because of the fuzziness and uncertainty of subjective responses, Fuzzy Set T theory is applied to identify critical challenges. A total of 65 questionnaires were distributed and 36 questionnaires were returned. The results revealed that the critical challenges faced by an ArPM are "poor planning," "unfamiliar technology," "unfamiliarity with green buildings and materials," "inappropriate scheduling," and "poor workmanship." All critical challenges were then categorized into six main groups including technical, managerial, personal skills, contractual, psychological, and financial.

# INTRODUCTION

The construction industry inherently has many different problems and requirements [1]. The importance of taking measures to improve the performance of the construction industry at various levels of socioeconomic development has been recognized in many countries. While the construction industry everywhere faces problems and challenges, the complexities and difficulties are most critical in developing countries due to the general situation of socioeconomic issues, lack of resources, organizational weaknesses, and an inability to deal with critical situations. According to the Singapore Department of Building [2], there is evidence that the problem has become greater in extent and severity in recent years.

In Malaysia for instance, building construction accounts for about 67.6% of overall construction work and is considered to be an essential element of the industry [3]. Due to existing limitations such as unavailability or lack of technical and/or financial resources, Malaysia faces some challenges in local projects. Moreover, external factors, for example, prohibition to engage in commercial activities, political pressure to suddenly increase the volume of mass projects, and complex or difficult land conditions, may also complicate the problem [4].

Generally, there are some important elements for the successful management of a construction project such as project management philosophies, processes, methods, and tools; however, one important factor in this field is the capability of the project manager (PM). Finding the right PM for a construction project is the most important commercial implementation task. Selecting a capable PM is an essential element to achieve success in a construction project. However, PMs have a variety of personal characteristics based on their beliefs, experience, personal values, professional ethics, technical knowledge, superior social skills, and their management abilities [5]. In this regard, initial studies on architecture challenges were published during the 1960s [6]. According to Tan [7], most PMs come from a wide range of professional disciplines in developing countries. Consequently, not one person from a single profession can claim to be a perfect PM. In some countries, there are no guidelines or standards that provide for a better selection of PMs to ensure excellent performance in the construction industry. However, certain organizations such as the Project Management Institute (PMI) have submitted their Strategic Overview of Project Management practices to the Construction Industry Development Board (CIDB) and recommended general directions to improve the service for the benefits of the governmental sector [8].

Construction project abandonment has been frequently reported in many countries such as the United States [9], Spain [10], United Arab Emirates, Saudi Arabia, Qatar, Bahrain, Kuwait, and Russia [11]. Although the reducing trend shows satisfactory results, the Ministry of Housing and Local Government reported that the lack of having the competent PMs and mismanagement are still critical problems in Malaysian construction industry, such that 514 and 95 abandoned housing projects were reported in 2000 and 2012, respectively, [12–15]. From the records, it is shown that nearly 90% of abandoned projects were caused by management problems faced by the developer [15]. Only a few of the causes are due to technical matters and failure in the company's performance. The statistics also reveal that one of the main reasons for project failures in Malaysia and many other developing countries is the lack of a proper architect as a PM (ArPM), as an agreeable and competent manager with appropriate plans and point of view [16]. Malaysian standard form of building contract [17] enumerates some of the most important points that ArPM has to consider to face fewer challenges and be successful in construction projects including clear contract and communication, proper planning and monitoring, and also special skills to lead and handle problems at construction sites. This study specifically focuses on the identification of the most critical challenges that the architect may face as a PM in Malaysia as a developing country. Despite wide discussion in the literature regarding the importance of the role of PMs and architects in construction projects, few prior studies have identified the critical challenges facing an architect as a project manager.

## ARCHITECT AS A PROJECT MANAGER (ARPM)

Project management is the direction and supervision of a project by the use of specific tools and control techniques including cost control, manpower, time, plant/machineries, communication, and motivation [46]. According to Wideman [47], "project management is the art of directing and coordinating human and material resources to achieve stated objectives within limits of time, budget, and stakeholders' satisfaction." PMs have total responsibility in construction sectors such as planning, organizing, and controlling. They are responsible for making sure that the planning phase of a project contains complete task definition, resources, time schedule, and a list of requirements. Furthermore they organize the project to make sure of hiring competent staff in order to produce necessary services and to achieve project objectives.

Many studies have mentioned the critical prerequisite skills for a PM [48, 49]. In this regard, Ahadzie [50] had investigated the relationship between achieving project success and construction project management competencies. Successful construction organizations now focus on ensuring that PMs obtain the principal competencies that they require to be successful in their jobs. According to Frank [51], the PM has direct authority over 34–47% of a project's success. Accordingly, it is clear that a PM plays a key role in achieving success for a construction project. The relationship between client, developer, and/or investor and the rest of the project's construction team is also significant.

PMs play significant roles in the process of a construction/ rehabilitation project. A PM with adequate experience, supported by a group of qualified professional and clerical staff, will head the project and be responsible for the day-to-day implementation of project activities. A PM and his team should have training in management aspects [52]. PMs should be approved and accepted by project stakeholders to have adequate experience for heading the project implementation and be responsible for the supervision of specific implementation activities. In maintenance/retrofit projects for instance, various proposals for rehabilitation activities are presented to PMs in the form of project proposals. Accordingly, one challenge facing project managers and decision makers is to select the most vulnerable structures [53]. In this regard, the creation of alternatives for the renovation or rehabilitation of every project is mainly incumbent on the PMs [54]. Hegazy [55] also mentioned that it is the objective of PMs to optimize the allocation and leveling aspects of the project resources.

While PMs play important roles in the construction industry from the design stage to implementation and demolition, most employers and clients have paid little attention to the performance of the architects in construction projects. They would rather concentrate on the performance of contractors. Recently, the construction industry has integrated the architect as a manager with executive power on the site [56]. The performance of the architect is therefore important, because any decision made at the inception of the project will affect the project's success. In brief, although the construction industry faces many challenges such as delays, cost overruns, fluctuation in the price of materials, and breaches of duty, all of these challenges are directly related to the relationship between the ArPM and client.

# RESPONSIBILITIES AND SKILLS OF ARPM

According to Odusami [57], skill is defined as "an ability that can be developed which is manifested in performance." In other words, skills are the knowledge and abilities you gain throughout your life. Due to the quickly changing environment of the construction industry with emerging challenges such as skills shortages, the rapid advancement of information and communication technologies, and the increasing prioritization of issues such as sustainability, environmental protection, and climate change, the role of the PM needs to be adapted. Widespread research studies have documented that managerial skill is necessary for efficient project performance [57].

Generally, the construction manager has a challenging role to perform and must manage a wide variety of works on site. The project manager's CADD survival guide [58] describes the duties and responsibilities of PMs with the basic management components such as project control, project administration, financial control, client relations, business development, design production, team management, and quality assurance [58]. According to the Malaysian governmental form of contract [17], the construction PM has an obligation to serve the owner and also has right and authority to represent the owner and to carry out business on their behalf. Additionally, he/she has the right to choose contractors and manage the relationship between the contractor, owner, and staff.

All ArPMs have four main duties: leading, organizing, controlling, and supervising. Each function plays an essential role during the construction project. Indeed, ArPMs have to consider all steps from the design stage till the demolition. In the planning stage, they must act as a forecaster in order to predict and provide the following requirements: (i) establish aims and goals, (ii) organize activities to achieve goals, and (iii) monitor budget and time. Secondary skills are the organization of resources such as time, cost, manpower, and machinery and equipment to effectively approach the project aims. Another responsibility of the ArPMs is project control. Control processes measure progress against goals and evaluation activities and when to take corrective action. This skill ensures managers that all parts of the project are completed according to the contract, design, drawing, specification, and details of the architect. Finally, in the supervision step, all of the activities must be supervised during the project to prevent any defects in construction.

To standardize the ArPM required qualifications, the Project Management Institute (PMI) first documented nine knowledge areas: integration, time, cost, procurement, quality, communication, human resource, scope, and risk [56]. Each of the nine knowledge areas contains processes that need to be accomplished within its discipline in order to achieve an effective project management program. B. G. Fryer and M. Fryer [59] also listed social skills, decision-making skills, problem-handling skills, ability to recognize opportunities, and management of changes as key personal attributes affecting project success. It is explicitly clear that the role of today's PMs has changed. Some authors imply that PMs must supplement their traditional functions with non-engineering knowledge and skills in order to meet today's professional demands [60]. Today's PMs not only fulfill the traditional roles of project management but also must manage the project in the most efficient and effective manner with respect to sustainability. The former emphasizes workers' attributes such as knowledge, skills and abilities, and personal traits, while the latter treats work as existing independently of the worker, definable in terms of the technical requirements of the work [61].

Effective job performance is influenced by different characteristics of the competences such as behaviors and trait [62]. Crawford [49] provided an in depth understanding of the subject by proposing three classifications, namely, input competencies, personal competencies, and output competencies. He implied that input competencies refer to the knowledge and skills that a person brings to a job. In this regard, Majid et al. [46] categorized the PMs according to their skill levels into competent, excellent, and great. He also implied that a competent PM has the necessary knowledge, skill, and experience, an excellent PM has more skills beyond those individual characteristics, and a great PM has two additional skills, the motivation and natural talent to lead.

Consequently, the role and responsibilities of architects in the construction site are to handle the project efficiently and to overcome any challenges in its implementation. Failure on the part of the architect during the conceptual planning and design stage might lead to stress factors that cause significant problems in successive stages of the project. ArPM must have various required skills such as detail planning, scheduling and construction estimation of operating procedure, supervision, inspection and testing, plan and submittal review, property management, and correspondence. The workmanship of the ArPMs and their skills in terms of design and project management implementation are definitely one of the most significant factors in achieving the goals of a construction project. All of the abovementioned skills can be applied to motivate project employees, forecast conditions, control and supervise tasks, and manage resources.

# CHALLENGES OF ARPM AT CONSTRUCTION SITES

A construction PM should consider everything in order to prevent failure in any step of the project. However, in most cases, undesirable challenges exist in project implementation.

Generally, the implementation of any new system or a change in initiatives has always been a challenging task; some of these challenges can occur during the implementation or the practice stage of a project. Due to a lack of clarity surrounding program management in the construction industry, an understanding of the major challenges facing a project remains vague. Milosevic et al. [21] stated that a fair amount of confusion exists among organizations about what program management stands for, and there is a lack of sufficient literature to accurately describe it [20]. In addition, Young [23] mentioned that economic fluctuations, population and migration growth, and growing pressure from global economic instability are the general characteristics of the construction industry environment. Accordingly, program management is aimed at exchanging timely and useful information with stakeholders or work teams. Therefore, lack of cross-functional communication can be another major challenge for ArPM in managing a construction program.

Specifically, one of the significant challenges that architects are facing concerns their skills and ability to manage and handle problems, which has not been mentioned directly in previous studies. The ArPM has to be both skilled and knowledgeable in order to overcome challenges that they would encounter during the design stage or implementation stage. Lack of sufficient skills or knowledge will lead to poor communication that weakens the relationship between both PM and contractors and between PM and clients. Therefore, to avoid additional challenges on construction sites, a competent manager has to have the necessary skills to negotiate with clients and to create binding agreements. In brief, proper selection of a PM can allow the completion of a project within deadlines and a high standard of quality [63]. In terms of having the essential skills to overcome the challenges a PM may face in working closely with a range of other professionals, PMs organize, plan, schedule, and control the work of their subordinates. Moreover, completing the project within the time and cost constraints are the main responsibilities of the PMs [64].

Siva and London [38] implied that the relationship between employees and architect is important. According to Hewage et al. [65], over 45% of construction workers described communication problems between the managers and the workers; this inadequate communication clearly results in adverse impacts on the productivity of the workforce.

Construction PMs are generally considered to be the key in the success of a construction project. Their role includes not only planning, organizing, and supervising the project team and the progress of work on the construction project, but also overcoming and dealing with demanding time pressures, inherent uncertainties, and complexity in the structure. The stressful environment of a construction project may reduce the potential capability of the PM. According to a recent occupational stress study in the construction industry, nearly 70% of construction professionals suffer from stress, anxiety, or depression [3]. The impacts of stress on construction professionals should be seriously taken into consideration. An environment that causes stress ultimately causes challenges in the construction industry project implementation and resource management [18].

Apart from the specific challenges faced by ArPM, factors such as increases in the scope and complexity of the structure as well as the active systems for its infrastructure and buildings, integrated project delivery and building information modeling, inadequate design, and poor planning may also create additional challenges in the completion of a construction project [39].

According to Bertelsen et al. [66], an architect's incompetence in project management systems, especially at the briefing stage of the building process, may result in the poor coordination and supervision of the design team. Another challenge as described by Khoury [67] is the separation of the architectural design from systems design and engineering which has often occurred because of rapid changes in the design processes. The criticality of the performance of ArPM in relation to public and private sector is another issue. In this regard, Adinyira and Dafeamekpor [68] evaluated the clients' perception of the performance of architects in building projects. According to NCARB

[69], the involvement of the ArPMs should be continued until and even after the completion of the construction documents.

Based on a critical literature review of the important role of PMs and ArPMs in construction projects, which were discussed in the previous sections, it is found that many efforts have been made to highlight different aspects of PMs. Although existing studies have argued and emphasized the importance of selecting the appropriate PM in construction/rehabilitation projects, critical challenges facing an architect as PM are still unavailable which presents a barrier to the effective assessment of the capabilities of an ArPM.

Project management philosophies, processes, methods, and tools are important components in successful project management; however, one of the most important elements is the capability of the PM. Finding the right PM for a construction project is therefore a major task in project implementation. Every owner, consultant, and contractor are on the lookout for a good project manager [70].

In Malaysia, there are no specific guidelines and standards that ensure a better selection of PMs for excellent performance in the construction industry. Project managers, more than any other individual, exert an immense influence over a project and can turn a potential success into an abject failure and a potential failure into a resounding success. Despite their influence, an unsuccessful project does not always indicate an unsuccessful PM. There are many external factors that cause a project to fail [71]. Truly, an unsuccessful project does not rely on the fault of a PM. But, the question is "how well do these project managers perform in the current construction industry?"

Accordingly, this study was carried out to investigate the challenges facing architects as project manager which are an essential part of the industry. More specifically, the aim of this study is to address the challenges of an ArPM in construction industry in general and in Malaysia, as developing country used as case study. Accordingly, there are two objectives: first, to investigate the conventional challenges of an ArPM in the Malaysian construction industry and, second, to identify the most critical challenges that have the greatest influence on a project's outcome. Data for analyzing the second objective was collected through a questionnaire survey.

# RESEARCH METHODOLOGY

The first stage in the identification of the critical challenges facing an ArPM in Malaysia is to identify the most common and conventional challenges that exist in such projects in general. The identification process was implemented through the thorough study of official documents and existing feasibility study reports. The first stage led to the generation of a list of relative challenges that could be categorized into smaller groups. In addition, a few interviews and discussions (as a pilot study) provided valuable comments on the adequacy of the selection of the preliminary challenges, which were incorporated in the main questionnaire survey. In fact, the pilot survey conducted to test the adequacy and comprehensiveness of the identified challenges was done with the help of selected experts. This preliminary pilot survey with the help of member checking [72] approved the respondent validation through offering the respondents the opportunity to add further challenges or to put them on record to provide summaries and to check the adequacy of the survey.

Based on the generated list of challenges, a questionnaire survey was then conducted to collect data from the targeted respondents including Malaysian government agencies, building and project consultants, general contractors, and developers. In order to identify the critical challenges of an ArPM in Malaysian construction projects, the experts and professionals from selected organizations were targeted. The organizations include Malaysian government agencies, civil, structure, and building consultants, project management consultants, and general contractors and developers. Coverage of such a variety of organizations was planned to represent the local construction industry in a comprehensive manner. The questionnaires were distributed among engineers and professionals in construction industry that would encounter and be familiar with the critical challenges in implementing a project in order to boost the quality and quantity of response.

## Data Collection

The data for determining the level of significance of the identified challenges were collected through a questionnaire survey. A 5-point Likert scale was used to capture the preferences of respondents in the questionnaire for each identified challenge, where a level of "5" indicates the most important, "4" important, "3" average, "2" unimportant, and "1" negligible. In this survey, a group of 65 experts were asked separately to indicate an appropriate significance level from different Malaysian organizations. In total, 36 valid questionnaires were collected among different fields. Table 1 shows the profile of the respondents based on their type of organizations.

**Table 1:** Respondents profile.

| Organization | Government agencies | Landlords | General contactors | Project management consultants | Civil and structural consultants | Architectural consultancy |
|---|---|---|---|---|---|---|
| Number of respondents | 4 | 0 | 6 | 7 | 3 | 16 |

As the respondents were all selected from Malaysian construction industry and had knowledge of the research concerns, the responses were of good quality, and a high response rate was ensured. Among the responses, 4 were from government agencies, 6 were from general contractors, 7 were from project management consultants, 3 were from civil and structural consultants, and 16 were from architectural consultants.

## Data Validity

One of the most important considerations when carrying out a questionnaire survey is the identification and selection of potential members to constitute the panel of experts. The selection of members or respondents is essential because the validity of the study is directly related to this selection process. Validity is a key aspect of scientific accuracy to ensure measuring the right thing and doing so properly. Without validity, there is little assurance that data is both relevant and accurate [73]. The validity of the postal questionnaire in this study is seen from two viewpoints: first, whether respondents who complete questionnaires do so accurately, honestly, and correctly and, second, whether those who fail to return their questionnaires would have given the same distribution of answers as did the returnees [74]. The question of accuracy can be checked by means of the interview method [75].

In the current study, therefore, validity was checked through the feedback process and structured interviews with a few selected professionals. One obvious advantage of this feedback process is its ongoing consideration of the goal assumptions' validity. In total, thirty-six effective responses were received with a response rate of 55%. According to Cohen et al. [75], a well-planned postal survey should obtain at least a 40-percent response rate without the judicious use of reminders. Therefore, the validity of the survey was seen. It was decided to categorize the respondents who have a direct interest and participation in the design and development of construction sites. The architectural engineer group includes the design professionals with a high rate of training and experience in design and an appreciation of the construction industry, particularly architecture. The professionals were randomly chosen and consisted of those who have had formal experience or training in any construction management process. The field distribution according to the type of organization is shown in Figure 1.

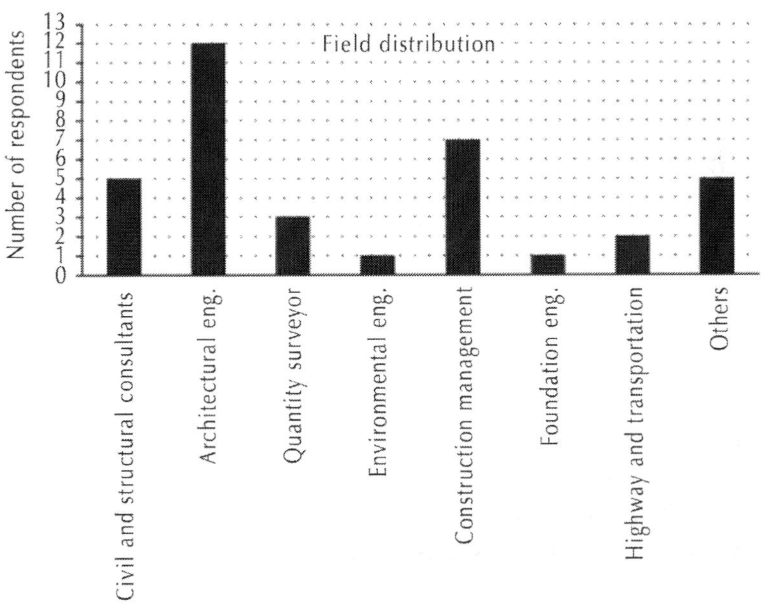


**Figure 1:** Type of work carried out by respondent's organizations (# of respondents: 36).

## Identification of Critical Challenges Based on Fuzzy Set Theory

The concept of fuzziness and the application of Fuzzy Set Theroy in data analysis were first introduced by Bellman and Zadeh [76] to deal with subjectivity and uncertainties. This is a strong and widely applicable theory for many fields of science and design including engineering, agriculture, medicine, and social science. This theory helps to identify the critical elements amongst all identified challenges. The main reason to apply this theory is because of fuzziness and uncertainty in subjective responses due to imprecision or vagueness [77]. In the current study, because of subjective and qualitative terms of the data in determining the significance of challenges, implementing this theory is beneficial [78].

Statistical calculations were conducted to determine the significance of critical challenges in the construction industry. Mean scores (MS) and standard deviations (SD) are determined for all challenges. MS and SD provided the basic data to which Fuzzy Set Theory was applied [79]. In the following paragraphs, the practical basis of this theory is described. For more details about the mathematical fundamentals of Fuzzy Set Theory, see Shen et al. [77] and Xia et al. [78]. According to Fuzzy Set Theory, the degree of membership is the probability of a variable belonging to a group [79] and is defined as follows [77]:

$$\mu\vec{A}(x_i) = \int_5^{\infty} F(Sx_i)d_x = 1 - Pf \tag{1}$$

Where $\mu\vec{A}(x_i)$ is the degree of membership for each challenge in each category and Z is the possibility that a challenge does not belong to the group of critical ones. The parameter is defined as the calculated value for whether or not a challenge is included in final set. Figure 2 shows the graphical illustration of the Z parameter. This variable is determined for each challenge with the help of the following equation:

$$Z = \frac{\text{Mean} - 2.5}{\text{SD}} \qquad (2)$$

The degree of membership for the respondents is estimated as follows:

$$\mu_A(X_i) = \min\{1, \mu_A(x)\} \qquad (3)$$

In this equation, $\mu_A(X_i)$ are the degrees of membership for each challenge. To identify critical challenges amongst the other challenges, the -cut set approach is adopted [77]. The $\lambda$-cut set method can be used to transfer a fuzzy set to a classical set. The optimal outcome is $\lambda$ =1 and the worst outcome is $\lambda$=0. For this study, $\lambda$=0.85 was adopted in order to identify the critical challenges [77, 80]. Figure 3 illustrates the flowchart for the procedure used to identify the critical challenges.

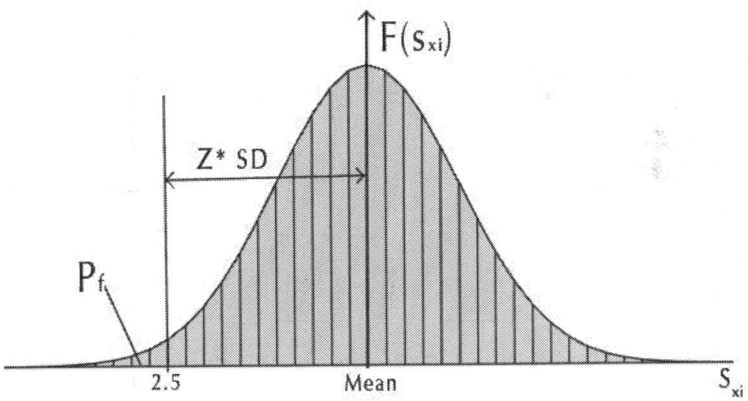


**Figure 2:** Normal distribution of challenges' significance score.

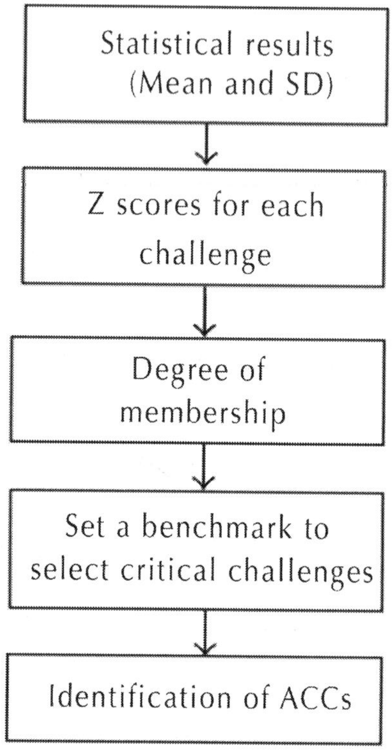


**Figure 3:** The Fuzzy Set Theory procedure to identify ACCs.

# RESULTS AND DISCUSSION

This section describes the findings of analysis regarding the most critical factors that the Malaysian ArPMs encounter in both the evaluated literature and their projects. During the first stage, a comprehensive literature review for finding the challenges was carried out. This section led to the introduction and identification of all ArPM challenges that had the greatest effects on goals of a project such as time, cost, and quality. 41 challenges were identified in the first stage. Table 2 shows the results of identifed challenges and their relative codes and descriptions for further analysis. It is to be noted that the identified challenges were revised and combined where possible.

**Table 2:** Summary of literature review regarding the identification of ArPM challenges.

| Code | Challenges | CIOB, 2006 [3]; Leung et al., 2008 [18] | Leung et al., 2008 [18] | Liao, 2007 [19] | Shehu and Akintoye, 2010 [20] | Milosevic et al., 2007 [21]; Cicmil et al. 2006 [22] | Young, CIOB, 2008 [23] | UK Construction Industry, 2005 [24] | Choi, 2008 [25] | OGC, 2007 [26] | Bartlett, 2002 [27]; Thiry, 2004 [28] | Reiss, 2003 [29] | Hwang and Ng, 2013 [30] | Tagaza and Wilson, 2004 [31]; Zhang et al., 2011 [32] | Chua et al., 2003 [33] | Eisenberg et al., 2002 [34] | Yitmen, 2007 [35] | Sexton and Barret 2004 [36] | Hewage et al., 2008 [37] | Siva and London, 2011 [38] | Tatum, ASCE, 2012 [39] | Hellmund et al., 2008 [40] | Mills and Glass, 2009 [41] | Bibby et al., 2003, [42]; Gorse and Emmitt, 2007 [43] | Hwang et al., 2009 [44]; Taib, 2010 [45] |
|---|---|---|---|---|---|---|---|---|---|---|---|---|---|---|---|---|---|---|---|---|---|---|---|---|---|
| CH1 | Stressful tasks | * | | | | | | | | | | | | | | | | | | | | | | | |
| CH2 | Poor planning | | * | * | * | * | | | | | | | | * | * | * | * | | | | * | | | * | |
| CH3 | Improper organizing | | * | | | | | | | | | | | | | | | | | | | | | | |
| CH4 | Lack of supervision | | * | | | | | | | | | | | | | | | | | | | | | | |
| CH5 | High number of tasks | | * | | | | | | | | | | | | | | | | | | | | | | |
| CH6 | Poor communication | | | * | | | | | | * | | | | | * | | * | * | * | * | | * | | * | |
| CH7 | Unsafe condition | | | * | | | | | | | | | | | | | | | | | | | | | |
| CH8 | Poor decision making | | | * | | | | | | | | | | | * | | | | | | | | | | |
| CH9 | Lack of clarity in program management | | | | * | | | | | | | | | | | | | | | | | | | | |
| CH10 | Lack of commitment from business leaders | | | | * | | | | | | | | | | | | | | | | | | | | |
| CH11 | Delay in project delivery | | | | * | * | | | | | | | | | | | | | | | | | | | |
| CH12 | Lack of cross-functional working | | | | * | * | | | | | | | | | | | | | | | | | | | |
| CH13 | Lack of coordination between tasks | | | | * | * | | | | | | | | | | | | | | | | | | | |
| CH14 | Lack of alignment between project and strategy | | | | * | * | | | | | | | | | | | | | | | | | | | |
| CH15 | Economic fluctuation | | | | * | | * | | | | | | | | | | | | | | | | | | |
| CH16 | Time pressure | | | | | | | * | | | | | | | | | * | * | | | * | * | | * | |
| CH17 | Lack of portfolio management knowledge | | | | | | | | * | | | | | | | | | | | | | | | | |
| CH18 | Lack of resources | | | | | * | | | | | * | | | | | | | | | | | | | | |
| CH19 | Improper relevant training | | | | * | | | | | | | * | | | | | | | | | | | | | |
| CH20 | Unrelated tasks | | | | | | | | | | | | * | * | * | | | | | * | | | | * | |
| CH21 | Complexities in construction process | | | | * | | | | | | | | | | * | | | | | | * | | | | |
| CH22 | Conflicting project objectives | | | | * | | | | | | | | | | | | | | * | | | | | | |
| CH23 | Fuzzy companies strategy | | | | * | | | | | | | | | | | | | | | | | | | | |
| CH24 | Discrepancy in contract forms | | | | | | | | | | | | | * | * | | | | | | | | | | |
| CH25 | Unfamiliar technology | | | | | | | | | | | | | * | | * | | | * | | * | | | | |
| CH26 | Unfamiliarity with Green building and materials | | | | | | | | | | | | | * | | * | | | | | | | | | |
| CH27 | Inappropriate construction sequence | | | | | | | | | | | | | | * | | * | * | | | | | | * | |
| CH28 | Difficulty in the selection of subcontractor | | | | | | | | | | | | | | * | | | | | | | | | | |
| CH29 | Design alteration | | | | | | | | | | | | | | * | | | | | | | | | | |
| CH30 | High level of risk | | | | | | | | | | | | | | * | | | | | | | | | | |
| CH31 | Worker's unawareness | | | | | | | | | | | | | | * | | | | * | | | | | | |
| CH32 | Government policy | | | | | | | | | | | | | | * | | | | | | | | | | |
| CH33 | Unforeseen circumstances | | | | | | | | | | | | | | * | | | | | | | | | | |
| CH34 | Lack of trust | | | | | | | | | | | | | | | | * | * | | * | | | | | |
| CH35 | Unclear contract | | | | | | | | | | | | | | | | * | * | | | | | * | * | |
| CH36 | Inappropriate performance measurement | | | | | | | | | | | | | | | | * | | | | | | | | |
| CH37 | Nonintegrated project delivery | | | | | | | | | | | | | | | | | | * | | * | * | * | * | |
| CH38 | Poor specification | | | | | | | | | | | | | | | | | | | | | | | | * |
| CH39 | Inappropriate scheduling | | | | | | | | | | | | | | | | | | | | | | | | * |
| CH40 | Poor workmanship | | | | | | | | | | | | | | | | | | | | | | | | * |
| CH41 | Inadequate payment | | | | | | | | | | | | | | | | | | | | | | | | * |

As previously mentioned, the data used for this study was collected through a questionnaire survey from targeted Malaysian experts. The adequacy and readability of the questionnaire were also tested with a pilot study. Five experts were involved in the pilot study, and their comments were incorporated into the final questionnaire. In responding to the questionnaire, respondents were invited to indicate the level of significance of each assessment indicator for indicating the most critical challanges faced by architects in a project by assigning a score between 1 and 5. However, the fuzziness involved in the experts' opinions should be considered. Fuzzy Set Theory is a practical approach in this regard to capture the fuzziness in the questionnaire survey date; the calculations are detailed in Table 3.

**Table 3:** Statistical analysis to identify ACCs based on Fuzzy Set Theory.

| Code | MS | SD | scores | | ACCs |
|---|---|---|---|---|---|
| CH1 | 1.40 | 0.55 | −2.01 | 0.02 | — |
| CH2 | 3.80 | 0.45 | 2.91 | 1.00 | ✓ |
| CH3 | 2.40 | 0.55 | −0.18 | 0.43 | — |
| CH4 | 2.60 | 0.55 | 0.18 | 0.57 | — |
| CH5 | 3.20 | 0.45 | 1.57 | 0.94 | ✓ |
| CH6 | 4.23 | 0.87 | 1.99 | 0.98 | ✓ |
| CH7 | 2.60 | 0.89 | 0.11 | 0.54 | — |
| CH8 | 3.80 | 0.84 | 1.55 | 0.94 | ✓ |
| CH9 | 3.20 | 0.84 | 0.84 | 0.80 | — |
| CH10 | 1.60 | 0.55 | −1.64 | 0.05 | — |
| CH11 | 3.40 | 0.55 | 1.64 | 0.95 | ✓ |
| CH12 | 3.80 | 0.84 | 1.55 | 0.94 | ✓ |
| CH13 | 2.80 | 0.45 | 0.67 | 0.75 | — |
| CH14 | 2.40 | 0.55 | −0.18 | 0.43 | — |
| CH15 | 2.00 | 0.71 | −0.71 | 0.24 | — |

| CH16 | 4.20 | 0.84 | 2.03 | 0.98 | ✓ |
|---|---|---|---|---|---|
| CH17 | 4.20 | 0.84 | 2.03 | 0.98 | ✓ |
| CH18 | 4.05 | 0.72 | 2.14 | 0.98 | ✓ |
| CH19 | 2.80 | 0.45 | 0.67 | 0.75 | — |
| CH20 | 4.09 | 0.97 | 1.64 | 0.95 | ✓ |
| CH21 | 2.00 | 0.71 | −0.71 | 0.24 | — |
| CH22 | 1.40 | 0.55 | −2.01 | 0.02 | — |
| CH23 | 3.00 | 0.71 | 0.71 | 0.76 | — |
| CH24 | 3.82 | 1.10 | 1.20 | 0.89 | ✓ |
| CH25 | 4.40 | 0.55 | 3.47 | 1.00 | ✓ |
| CH26 | 4.60 | 0.55 | 3.83 | 1.00 | ✓ |
| CH27 | 3.82 | 1.10 | 1.20 | 0.89 | ✓ |
| CH28 | 2.80 | 0.84 | 0.36 | 0.64 | — |
| CH29 | 4.09 | 0.75 | 2.12 | 0.98 | ✓ |
| CH30 | 3.73 | 0.77 | 1.60 | 0.95 | ✓ |
| CH31 | 3.95 | 0.95 | 1.53 | 0.94 | ✓ |
| CH32 | 1.20 | 0.45 | −2.91 | 0.00 | — |
| CH33 | 1.60 | 0.55 | −1.64 | 0.05 | — |
| CH34 | 2.60 | 0.55 | 0.18 | 0.57 | — |
| CH35 | 4.14 | 0.83 | 1.96 | 0.98 | ✓ |
| CH36 | 3.40 | 0.55 | 1.64 | 0.95 | ✓ |
| CH37 | 2.60 | 0.55 | 0.18 | 0.57 | — |
| CH38 | 4.27 | 0.77 | 2.31 | 0.99 | ✓ |
| CH39 | 4.32 | 0.65 | 2.81 | 1.00 | ✓ |
| CH40 | 4.41 | 0.50 | 3.79 | 1.00 | ✓ |
| CH41 | 4.18 | 1.01 | 1.67 | 0.95 | ✓ |

Architects have a variety of responsibilities within the construction and maintenance process. The questions in the survey were rated by the respondents to identify which challenges have the greatest influence on the projects outcome with respect to the perception of time, cost, and quality. These are the main parameters in planning and managing a successful construction project [81]. Inappropriate identification and selection of an ArPM may lead to poor planning which directly affect determining, analyzing, devising, and organizing the project resources. Subsequently, high alterations occurred in project outcomes in terms of time, cost, and quality. Therefore, the findings in this section can serve as a useful reference for construction practitioners and project managers, especially the architects, to revise the PM selection process and to avoid project challenges due to inappropriate selection of the PM.

The results reveal that the majority of the respondents believe that critical problems facing architects were "poor planning," "unfamiliar technology," "unfamiliarity with green buildings and materials," "inappropriate scheduling," and "poor workmanship" with $\mu(X_i)=1.00$ having the highest influence on project time, cost, and quality. Specific benefits that can be achieved from the identification of ACCs are a reduction in the time and cost of the project. Even though proper training, hiring, and managing architecture professionals may not prevent project delays and excessive overruns in terms of cost and time, such factors can be mitigated by construction projects stakeholders giving attention to the challenges facing the architect before the project even begins. 23 critical challenges were identified and ranked by the degree of membership in descending order which can be categorized in six general groups. These groups are discussed in more detail in the following sections. The findings can be useful for government construction legislators to revise the project manager regulations in order to reduce the number of failures in abandoned construction projects and save huge amount of budget in Malaysia.

## Category 1: Technical

Three critical technical challenges are addressed in this category (see Table 4), including challenges associated with Unfamiliar Technology $\lambda=1.00$ and also Complexities in Construction Process ($\lambda=0.95$). Technical challenge is defined by Katz [82] as lack of understanding and proficiency in a specific kind of activity, particularly one involving methods, processes, procedures, or techniques. It involves specialized knowledge, analytical ability within a specialty, and facility to use the tools and techniques of a specific discipline. A successful PM is one who possesses technical competence which has been gained through a career in an advanced technological environment. Furthermore, many researchers have mentioned technical challenges as the most important group of activities that can be challenging for an architect as a project manager [19, 38, 42,43], not to mention the fact that technical challenges is also one of the most critical challenges that architects may encounter and can cause the failure of a project [18, 32, 39].

**Table 4:** Category 1: technical challenges

| Code | Challenge | Degree of membership |
|---|---|---|
| CH25 | Unfamiliar technology | 1.00 |
| CH20 | Unrelated tasks | 0.95 |

## Category 2: Managerial

The challenges relevant to the management area consist of Inappropriate Scheduling ($\lambda$=1.00), Poor Specification ($\lambda$=0.99), Design Alterations ($\lambda$=0.98), Lack of Portfolio Management Knowledge ,Delays in Project Delivery ($\lambda$=0.95), Inappropriate Performance Measurements ($\lambda$=0.95), High Number of Tasks $\lambda$=0.94, and also Inappropriate Construction Sequences ($\lambda$=0.89). The degree of membership for CH39 (Inappropriate Scheduling) is 1.00; this illustrates the most critical factor in this category (see Table 5).

**Table 5:** Category 2: managerial challenges

| Code | Challenge | Degree of membership |
|---|---|---|
| CH31 | Inappropriate scheduling | 1.00 |

| | | |
|---|---|---|
| CH38 | Poor specification | 0.99 |
| CH29 | Design alteration | 0.98 |
| CH17 | Lack of portfolio management knowledge | 0.98 |
| CH11 | Delay in project delivery | 0.95 |
| CH36 | Inappropriate performance measurement | 0.95 |
| CH05 | High number of tasks | 0.94 |
| CH27 | Inappropriate construction sequence | 0.89 |

## Category 3: Personal Skills

Category 3 illustrates the critical factors that are categorized as personal skill categories (see Table 6), including Poor Workmanship $\lambda = 1.00$, Poor Planning $\lambda = 100$, Unfamiliarity with Green Buildings and Materials $\lambda = 1.00$, Poor Communication $\lambda = 0.98$, Poor Decision-Making $\lambda = 0.94$, Worker's Unawareness $\lambda = 0.94$, and Lack of Cross-Functional Working $\lambda = 0.94$.

**Table 6:** Category 3: personal skill challenges

| Codee | Challenge | Degree of membership |
|---|---|---|
| CH40 | Poor workmanship | 1.00 |
| CH2 | Poor planning | 1.00 |
| CH26 | Unfamiliarity with green buildings and materials | 1.00 |
| CH6 | Poor communication | 0.98 |
| CH8 | Poor decision-making | 0.94 |
| CH31 | Worker's unawareness | 0.94 |
| CH12 | Lack of cross-functional work | 0.94 |

Personal characteristic and skills can be defined as the unique combination of the psychological traits we use to describe a person's workmanship, often called personality. Personality is defined as "the sum total of the physical, mental, emotional, and social characteristics of an individual" [83]. Disabilities in translating knowledge into action can be a great challenge. In others words, skills are the knowledge and abilities you gain throughout life. Some skills come almost naturally, while others come only by complete devotion to study and practice [84]. The degree of membership for CH40, CH2, and CH26 is 1.00, which are the most critical factors.

## Category 4: Contractual

Standard procedures and policies refer to the project manager's awareness of the recognition of the need for all aspects of an organization's operations to be consciously addressed and set down as a corporate statement in line with recognized standards, supported by systems appropriate to the achievement of the policy [16]. A procedure and contract are a series of related steps or tasks expressed in chronological order for a specific purpose. Contract and procedures define in a step-by-step fashion the methods through which policies are achieved. Contract emphasizes detail and policies which differ from the procedures because the purpose of policies is to help in the objective setting process. Therefore, contract and policies also create an understanding among members of a group that make the actions of each member more predictable to other members [33].

The project manager can rely to some extent on the policies and procedures developed by the organization. Thus, a lack of understanding or a lack of clarity in the contract could cause serious problems related to the organization and management of the tasks given to contractors [20, 33]. Regarding the contractual challenges (see Table 7), two factors are identified as critical challenges including Unclear Contract $\lambda = 0.98$ and Discrepancy in Contract Forms $\lambda = 0.89$.

**Table 7:** Category 4: contractual challenges.

| Code | Challenge | Degree of membership |
|---|---|---|
| CH35 | Unclear contract | 0.98 |
| CH24 | Discrepancy in contract forms | 0.89 |

## Category 5: Psychological

There are two critical psychological challenges (see Table 8) which were identified in this category, including challenges relevant to Time Pressure $\lambda = 0.98$ and also High Level of Risk $\lambda = 0.95$. Psychological challenges are defined as the total activities, behaviors, and tasks that could be the cause of a stressor. Consequently, the stressor leads to the psychological stress of a person which disturbs activities of the organization [20, 31–33].

Table 8: Category 5: psychological challenges.

| Code | Challenge | Degree of membership |
|---|---|---|
| CH18 | Lack of resources | 0.98 |
| CH41 | Inadequate payment | 0.95 |

## Category 6: Financial

A lack of resources and inadequate finance from the client and payments for completed work during the construction stage are two common challenges found in most construction projects. In order for a project to be run efficiently and effectively by an architect as a project manager, management must provide sufficient resources in terms of labor, time, and money. Project management training shows how to define project needs and how to obtain approval upfront and helps project managers assign and prioritize resources throughout the duration of a project. To improve the chances of success from day one, an architect as project manager needs to be familiar with (i) how to approach their bank knowing exactly what they need in terms of finance, (ii) marketing their offering, (iii) managing their cash flow, preferably with a proper accounting system, and (iv) working on generating a spread of small and big contractors, rather than focusing on one or two large ones.

Thus, in order to manage a whole project in terms of contractors, materials, time, and cost, the architecture has to control and manage the finances to prevent such challenging matters in construction sites such as delays, disputes, and dissatisfaction in the relationships with the landlord and contractors. Accordingly, competent architects have to mention and include these kinds of issues in their design, test, inspection, and executive managing on site in order to lead the project and accomplish its goals. Lack of Resources ($\lambda = 0.98$) and Inadequate Payment ($\lambda = 0.95$) were identified as the two critical challenges (see Table 9).

**Table 9:** Category 6: financial challenges.

| Code | Challenge | Degree of membership |
|---|---|---|
| CH18 | Lack of resources | 0.98 |
| CH41 | Inadequate payment | 0.95 |

# CONCLUSION

The understanding of architecture framework as project manager by public, organizational, and governmental sectors is significant in terms of reliable and effective adoption among clients. The purpose of this study is to help residential building project practitioners in the professional sectors to better choose project managers, provide information technologies that create progressive processes, improve decision-making, and create a competitive advantage from its adoption.

More specifically, this paper investigates the criticality of challenges that architects must tackle in construction projects by applying practical methods to analyze the data collected from questionnaire surveys. A group of 41 different challenges were identified through a review of the literature. Malaysian experts were then employed to revise the data and prioritize the criticality of the challenges according to their experiences. Because of the vagueness and impreciseness in the respondents' opinions, Fuzzy Set Theory was then applied to identify the most critical challenges for the Malaysian ArPMs. 23 critical challenges in six main groups were identified according to their degree of membership. These groups were technical, personal, managerial skills, contractual, psychological, and financial. In addition, the top six ACCs are identified as "unfamiliarity with green building and materials," "poor workmanship," "unfamiliar technology," "poor planning," "inappropriate scheduling," and "poor specification" with degree of membership of one.

Since it could be viewed in various perspectives, the authors of this paper highlight the importance of integrating the existing challenges and applications in the professional sector organizations in order to establish an efficient framework for landlords and architects. Therefore, support for essential skills and knowledge, its applications, and the necessary infrastructure technologies can be used to create a reliable and flexible method for construction projects to access competent architects by using government data and information transfers within and between organizations to aid in their identification and classification and support their control over a project as well as the necessary characteristics and managerial skills to lead a project to success. By using the identified weighted challenges, the architect framework illustrates the infrastructure components, applications, and technologies that can be used as guidelines for project adoption. However, the limitations of the method are appreciated such as the limited number of respondents, which requires a larger population of experts, and the appropriate selection of -set which has a direct effect on the critical challenges which need to be rectified. It is also suggested that in order to reduce or even completely solve the challenges facing ArPMs, further studies be conducted on the identification of the personal characteristics, skills, roles, and responsibilities in construction projects required by an ArPM. This can be facilitated by the development of software applications for the evaluation of ArPMs.

# APPENDIX

## Definition of the ACCs

CH2: Poor Planning. It involves the lack of having choice of technology, the definition of work tasks, the estimation of the required resources and durations for individual tasks, and the identification of any interactions among the different work tasks.

CH5: High Number of Tasks. High project task dependencies mean that certain tasks cannot begin unless another is completed. To handle such projects individually is difficult and sometimes impossible. The project should be divided among some different PMs.

CH6: Poor Communication. There are problems concerning communication with a focus on intra supplier communication within the construction sector, demand-supply communication during the design phase, and communication between and within single demand and supply side parties, during the whole of the construction process [85].

CH8: Poor Decision-Making. It includes poor project governance, the lack of accountability of project stakeholders, and a lack of leadership.

CH11: Delay in Project Delivery. Delay is the time during which some part of the construction project has been extended or not performed due to an unanticipated circumstance [86].

CH12: Lack of Cross-Functional Working. Insufficient or inappropriate team working activities, functional diversity, competing identities, performance expectation, synergy, and integration with the construction organizations' structure will affect the future organizational strategy and development of the project. The PM is the leader of a cross-functional group comprising the project team assembled to successfully execute a set of project objectives.

CH16: Time Pressure. A limitation period is the period of time within which a party to a contract must bring a claim. In construction contracts, limitation periods are often relevant in relation to defects claims brought against contractors [3].

CH17: Lack of Portfolio Management Knowledge. There is lack of project managers with the fundamental practices needed to achieve organizational results and excellence in the practice of project management [56].

CH18: Lack of Resources. There is shortage in having enough workers, money, machinery, time, and so on.

CH20: Unrelated Tasks. They are the responsibilities which are not relevant, pertinent, or associated with the architect and are asked of him/her to successfully manage simultaneously.

CH24: Discrepancy in Contract Forms. It is any conflict between tasks and written agreement that occurs during the construction [16].

CH25: Unfamiliar Technology. It is specified technology which the architect has never encountered before and should retain its associated risks; including unexpected design problems, uncertain risks, and overspending on the project.

CH26: Unfamiliarity with Green Building and Materials. It is architect's awareness about materials and processes that is environmentally responsible and resource-efficient throughout a building's life-cycle: design, construction, operation, maintenance, renovation, and demolition.

CH27: Inappropriate Construction Sequence. It is improper describing of the sequence of the activities and the significant tasks interdependencies required to meet the project scope of work.

CH29: Design Alteration. It is any conflict, differences, or changes of design during the period of construction that may change basic tasks [16].

CH30: High Level of Risk. It is a project critical strategic importance and large budget and high complexity.

CH31: Worker's Unawareness. It is employees with low level of training and awareness in terms of construction tasks.

CH32: Unclear Contract. It is greement with any irrelevant or fuzzy clause, condition, and specification that may cause some challenges during the construction tasks.

CH36: Inappropriate Performance Measurement. It is lack of knowledge in assessment value procedure of essential performance parameters of the current project through engineering analysis and tests.

CH38: Poor Specification. It is lack of specifications to clarify and advise the contractor of his responsibility to perform all of the work and services as outlined in construction tasks Department of Public Work, Dartmouth, Massachusetts, Construction Specifications, 2009,http://www.town.dartmouth.ma.us/Pages/DartmouthMA_DPW/Con.pdf.

CH39: Inappropriate Scheduling. It is ineffective, unrealistic, and impossible project time estimation for project's milestones, activities, and deliverables.

CH40: Poor Workmanship. The Oxford Dictionary defines it as "the relative finish or execution seen in the manufactured article." Poor workmanship has been cited as the major cause of leakage problems in construction sites [87].

CH41: Inadequate Payment. It is all about payment with delay or regardless of contract that causes dispute between construction staff and employer.

## CONFLICT OF INTERESTS

The authors declare that there is no conflict of interests regarding the publication of this paper.

## ACKNOWLEDGMENT

The authors would like to express their heartfelt gratitude to Dr. Hossein Moayedi, the editor, and the anonymous reviewers who have enhanced the contents of the present study significantly with their insightful comments. This work was financially supported by the eScienceFund (Project no. 06-01-06-SF1198), Ministry of Science, Technology and Innovation of Malaysia (MOSTI), UTM Construction Research Centre, and Research Management Centre.

## REFERENCES

1. P. M. Hillebrandt, "Problems of larger local contractors: causes and possible remedies," CIB REPORT, 1999.
2. National University of Singapore Department of Building, 2010,
3. CIOB, "Occupational stress in the construction industry," CIOB national stress survey results, 2006,
4. A.-R. Abdul-Aziz and P. S. Jahn Kassim, "Objectives, success and failure factors of housing public-private partnerships in Malaysia," Habitat International, vol. 35, no. 1, pp. 150–157, 2011.

5. A. Majid, M. Zaimi, and R. Zakaria, An effectiveness of communication in project undertaking: problems and constraints, 2001.
6. I. Avots, "Why does project management fail? (Project management systems failure analysis, discussing cost, products quality and project objectives)," California Management Review, vol. 12, pp. 77–82, 1969.
7. A. A. L. Tan, People Skills in Project Management, Venton Publishing (M) Sdn. Bhd., Batu Caves, Malaysia, 2006.
8. F. B. Razalli, Roles, skills and characteristics of project manger in construction industry [M.S. thesis], Faculty of Civil Engineering, Universiti Teknologi Malaysia, May 2007.
9. M. Hicks, "United States, Real Estate, Contract, Project Abandonment: Separate Legal Concepts," Mondaq, 2008,
10. R. Carrero, G. Malvárez, F. Navas, and M. Tejada, "Negative impacts of abandoned urbanisation projects in the Spanish coast and its regulation in the Law," Journal of Coastal Research, pp. 1120–1124, 2009.
11. SPIEGEL, Architecture's Reality Check: Global Downturn Dooms Prestige Construction Projects, SPIEGEL ONLINE International, 2011,
12. A. Kabit, "Woes that cannot be abandoned," The Star Online, 2010,
13. L. M. A. Bettencourt and J. Kaur, "Evolution and structure of sustainability science," Proceedings of the National Academy of Sciences of the United States of America, vol. 108, no. 49, pp. 19540–19545, 2011.
14. N. Heng, "Abandoned housing projects revisited," Sun2Surf, 2011,
15. MHLG (Ministry of Housing and Local Government), Perangkaan Terpilih KPKT Sehingga, Ministry of Housing and Local Government, 2012,
16. S. Rajoo, "The PAM 2006 standard form of building contract—a change in risk allocation," Malayan Law Journal, 2010.
17. P.W.D Form 203. Standard form of contract to be used where drawings and specification form part of the contract, Rev., 2007.

18. M. Leung, Y. Chan, and P. Olomolaiye, "Impact of stress on the performance of construction project managers," Journal of Construction Engineering and Management, vol. 134, no. 8, pp. 644–652, 2008.
19. S. S. C. Liao, "Top ten qualities of project manager Heroes: an interview with David B. Perini,"Leadership and Management in Engineering, vol. 7, no. 2, pp. 50–60, 2007.
20. Z. Shehu and A. Akintoye, "Major challenges to the successful implementation and practice of programme management in the construction environment: a critical analysis," International Journal of Project Management, vol. 28, no. 1, pp. 26–39, 2010.
21. D. Z. Milosevic, R. J. Martinelli, and J. M. Waddell, Program Management for Improved Business Results, John Wiley & Sons, 2007.
22. S. Cicmil, T. Williams, J. Thomas, and D. Hodgson, "Rethinking project management: researching the actuality of projects," International Journal of Project Management, vol. 24, no. 8, pp. 675–686, 2006.
23. N. W. Young, "Key trends in the European and US construction market place," Smart Market Report, Design and Construction intelligence, CIOB, University of Reading and CRC, 2008.
24. UK Construction Industry; Construction Industry News, "Industry Structure," 2005,
25. B. Choi, S. K. Poon, and J. G. Davis, "Effects of knowledge management strategy on organizational performance: a complementarity theory-based approach," Omega, vol. 36, no. 2, pp. 235–251, 2008.
26. OGC, Project management, 2007,
27. J. Bartlett, Managing Programmes of Business Change: A Handbook of the Principles of Programme Management, Project Manager Today Publications, 2002.
28. M. Thiry, ""For DAD": A programme management life-cycle process," International Journal of Project Management, vol. 22, no. 3, pp. 245–252, 2004.
29. G. Reiss, Programme Management Demystified: Managing Multiple Projects Successfully, Resources/Publications, Spon Press, 2003.

30. B.-G. Hwang and W. J. Ng, "Project management knowledge and skills for green construction: overcoming challenges," International Journal of Project Management, vol. 31, no. 2, pp. 272–284, 2013.
31. E. Tagaza and J. L. Wilson, "Green buildings: drivers and barriers e lessons learned from five Melbourne developments," Tech. Rep., Building Commission by University of Melbourne and Business Outlook and Evaluation, 2004.
32. X. Zhang, L. Shen, and Y. Wu, "Green strategy for gaining competitive advantage in housing development: a China study," Journal of Cleaner Production, vol. 19, no. 2-3, pp. 157–167, 2011.
33. D. K. H. Chua, Y. Wang, and W. T. Tan, "Impacts of obstacles in East Asian cross-border construction,"Journal of Construction Engineering and Management, vol. 129, no. 2, pp. 131–141, 2003.
34. D. Eisenberg, R. Done, and L. Ishida, "Breaking down the barriers: challenges and solutions to code approval of green building," Research Report, Development Center for Appropriate Technology, 2002,
35. I. Yitmen, "The challenge of change for innovation in construction: a North Cyprus perspective,"Building and Environment, vol. 42, no. 3, pp. 1319–1328, 2007.
36. M. Sexton and P. Barrett, "The role of technology transfer in innovation within small construction firms," Engineering, Construction and Architectural Management, vol. 11, no. 5, pp. 342–348, 2004.
37. K. N. Hewage, J. Y. Ruwanpura, and G. F. Jergeas, "IT usage in Alberta's building construction projects: current status and challenges," Automation in Construction, vol. 17, no. 8, pp. 940–947, 2008.
38. J. P. S. Siva and K. London, "Investigating the role of client learning for successful architect-client relationships on private single dwelling projects," Architectural Engineering and Design Management, vol. 7, no. 3, pp. 177–189, 2011.
39. C. B. Tatum, "Integrated construction engineering activities to satisfy challenging project objectives," inProceedings of the

Construction Research Congress, Construction Challenges in a Flat World, pp. 139–148, ASCE, 2012.

40. A. J. Hellmund, K. G. Van Den Wymelenberg, and K. Baker, "Facing the challenges of integrated design and project delivery," Strategic Planning for Energy and the Environment, vol. 28, no. 1, pp. 69–80, 2008.
41. F. T. Mills and J. Glass, "The construction design manager's role in delivering sustainable buildings,"Architectural Engineering and Design Management, vol. 5, no. 1-2, pp. 75–90, 2009.
42. L. Bibby, S. Austin, and D. Bouchlaghem, "Design management in practice: testing a training initiative to deliver tools and learning," Construction Innovation, vol. 3, no. 4, pp. 217–229, 2003.
43. C. A. Gorse and S. Emmitt, "Communication behaviour during management and design team meetings: a comparison of group interaction," Construction Management and Economics, vol. 25, no. 11, pp. 1195–1211, 2007.
44. B. Hwang, S. R. Thomas, C. T. Haas, and C. H. Caldas, "Measuring the impact of rework on construction cost performance," Journal of Construction Engineering and Management, vol. 135, no. 3, pp. 187–198, 2009.
45. T. Taib, Efficiency in construction process [Ph.D. thesis], Universiti Teknologi Malaysia, Faculty of Civil Engineering, 2010.
46. M. Z. A. Majid, W. Z. Zakaria, and A. Keyvanfar, "Executive information site management system for monitoring project performance: system requirement study," OIDA International Journal of Sustainable Development, vol. 3, no. 3, pp. 11–24, 2012.
47. R. M. Wideman, "Comparing PRINCE2 with PMBoK," 2002,
48. W. Belassi and O. I. Tukel, "A new framework for determining critical success/failure factors in projects," International Journal of Project Management, vol. 14, no. 3, pp. 141–151, 1996.
49. L. Crawford, "Profiling the competent project manager," in Proceedings of the PMI Research Conference, pp. 3–15, Project Management Institute, June 2000.
50. D. K. Ahadzie, A model for predicting the performance of project managers in mass house building projects in Ghana.

51. T. Frank, The Superior Project Manager, Marcel Dekker, New York, NY, USA, 2002.

52. M. A. Khan, Bridge and Highway Structure Rehabilitation and Repair, McGraw-Hill, 2010.

53. M. Yadollahi and M. Z. Zin, "Development of analytical hierarchy process (AHP) method for rehabilitation project ranking before disasters," in Proceedings of the 2nd International Conference on Disaster Management and Human Health: Reducing Risk, Improving Outcomes, pp. 209–220, WIT Press, Orlando, Fla, USA, 2012.

54. I. M. Shohet and E. Perelstein, "Decision support model for the allocation of resources in rehabilitation projects," Journal of Construction Engineering and Management, vol. 130, no. 2, pp. 249–257, 2004.

55. T. Hegazy, "Optimization of resource allocation and leveling using genetic algorithms," Journal of Construction Engineering and Management, vol. 125, no. 3, pp. 167–175, 1999.

56. PMBOK Guide, Project Management Body of Knowledge (PMBOK Guide), Project Management Institute, 2008.

57. K. T. Odusami, "Perceptions of construction professionals concerning important skills of effective project leaders," Journal of Management in Engineering, vol. 18, no. 2, pp. 61–67, 2002.

58. S. M. Benz, The Project Manager's CADD Survival Guide, ASCE Publications, 1997.

59. B. G. Fryer and M. Fryer, The Practice of Construction Management, BSP Professional, 1990.

60. T. Ceran and A. A. Dorman, "The complete project manager," Journal of Architectural Engineering, vol. 1, no. 2, pp. 67–72, 1995.

61. L. Holmes and P. Joyce, "Rescuing the useful concept of managerial competence: from outcomes back to process," Personnel Review, vol. 22, no. 6, pp. 37–52, 1993.

62. S. E. Abraham, L. A. Karns, K. Shaw, and M. A. Mena, "Managerial competencies and the managerial performance appraisal process," The Journal of Management Development, vol. 20, no. 10, pp. 842–852, 2001.

63. M. Z. A. Majid, W. Z. Zakaria, and A. Keyvanfar, "Executive information site management system for monitoring project performance: system requirement study," OIDA International Journal of Sustainable Development, vol. 3, no. 3, pp. 11–24, 2012.
64. S. K. Sears, G. A. Sears, and R. H. Clough, Construction Project Management: A Practical Guide to Field Construction Management, Wiley, Hoboken, NJ, USA, 5th edition, 2008.
65. K. N. Hewage, J. Y. Ruwanpura, and G. F. Jergeas, "IT usage in Alberta's building construction projects: current status and challenges," Automation in Construction, vol. 17, no. 8, pp. 940–947, 2008.
66. S. Bertelsen, K. Fuhr Petersen, and H. Davidsen, The Client as Agent for Changes—Towards a New Culture in Building, Bygherreforeningen Danmark, Fonden Realdania, Byggecentrum, Denmark, 2002.
67. K. B. Khoury, Communication in architecture project management: the case of architecture firms in Lebanon, [A thesis submitted in partial fulfillment of the requirements for the degree of Doctor of Business Administration], Grenoble Ecole de Management, 2012.
68. E. Adinyira and C. A. Dafeamekpor, "Clients' perceptions of architect performance on building projects in Ghana," of Professional Issues in Engineering Education & Practice, 2014.
69. Necessity of an Architect During Construction (NCARB), National Council of Architectural Registration Boards, Washington, DC, USA, 20006-1310, December 2010,
70. S. Ogunlana, Z. Siddiqui, S. Yisa, and P. Olomolaiye, "Factors and procedures used in matching project managers to construction projects in Bangkok," International Journal of Project Management, vol. 20, no. 5, pp. 385–400, 2002.
71. J. Wateridge, "How can IS/IT projects be measured for success?" International Journal of Project Management, vol. 16, no. 1, pp. 59–63, 1998.
72. Y. S. Lincoln and E. Guba, Naturalistic Inquiry, Sage, Beverly Hills, Calif, USA, 1985.

73. F. Adam and P. Humphreys, Encyclopedia of Decision Making and Decision Support Technologies, United States of America, Information Science Reference, 2008.

74. W. A. Belson, Validity in Survey Research, Gower, Aldershot, UK, 1986.

75. L. Cohen, L. Manion, and K. Morrison, Research Methods in Education, Routledge, Taylor & Francis, New York, NY, USA, 6th edition, 2007.

76. R. E. Bellman and L. A. Zadeh, "Decision-making in a fuzzy environment," Management Science, vol. 17, no. 4, pp. 141–164, 1970.

77. L. Shen, W. Lu, Y. Peng, and S. Jiang, "Critical assessment indicators for measuring benefits of rural infrastructure investment in China," Journal of Infrastructure Systems, vol. 17, no. 4, pp. 176–183, 2012.

78. B. Xia, A. P. C. Chan, and J. F. Y. Yeung, "Developing a fuzzy multicriteria decision-making model for selecting design-build operational variations," Journal of Construction Engineering and Management, vol. 137, no. 12, pp. 1176–1184, 2011.

79. H. J. Zimmermann, Fuzzy Set Theory and Its Applications, Kluwer Academic, London, UK, 4th edition, 2001.

80. T. Tervonen, J. R. Figueira, R. Lahdelma, J. A. Dias, and P. Salminen, "A stochastic method for robustness analysis in sorting problems," European Journal of Operational Research, vol. 192, no. 1, pp. 236–242, 2009.

81. P. A. Bowen, K. S. Cattel, K. A. Hall, P. J. Edwards, and R. G. Pearl, "Perceptions of time, cost and quality management on building projects," Australasian Journal of Construction Economics and Building, vol. 2, no. 2, pp. 48–56, 2012.

82. R. H. Katz, "Managing the chip design database," Computer, vol. 16, no. 12, pp. 26–36, 1983.

83. B. Sandoval, D. Mangus, and B. Samuels, Heritage Comics Auctions, 2005 Random House Archives Catalog# 816, Heritage Capital Corporation, 2005.

84. D. Clark, Big Dog's Leadership Page, 1997,

85. M. E. L. Hoezen, I. M. M. J. Reymen, and G. P. M. R. Dewulf, "The problem of communication in construction," in Proceedings of the CIB W06 Adaptables Conference, Eindhoven, The Netherlands, 2006.
86. M. T. Callahan, B. B. Bramble, and F. M. Rapoport, Discovery in Construction Litigation, Michie Company, 1987.
87. J. I. Davison, Workmanship and Rain Penetration of Masonry Walls, Atlantic Regional Station, Division of Building Research, National Research Council of Canada, 1980,

# Citations

## CHAPTER 1

R. U. Halwatura and N. P. N. P. Ranasinghe, "Causes of Variation Orders in Road Construction Projects in Sri Lanka," ISRN Construction Engineering, vol. 2013, Article ID 381670, 7 pages, 2013. doi:10.1155/2013/381670.

## CHAPTER 2

Hoon-ku Lee, Yoon-sun Lee and Jae-jun Kim (2010). A Cost-Based Interior Design Decision Support System for Large-Scale Housing Projects, Convergence and Hybrid Information Technologies, Marius Crisan (Ed.), ISBN: 978-953-307-068-1, InTech, DOI: 10.5772/9636. Available from: http://www.intechopen.com/books/convergence-and-hybrid-information-technologies/a-cost-based-interior-design-decision-support-system-for-large-scale-housing-projects

# CHAPTER 3

Jian Li, Ying Wang, Xiangyu Wang, Hanbin Luo , Shih-Chung Kang, Jun Wang, Jun Guo and Yi Jiao, Benefits of Building Information Modelling in the Project Lifecycle: Construction Projects in Asia DOI: 10.5772/58447

# CHAPTER 4

A.R. Rezaei, T. Çelik, Y. Baalousha, Performance measurement in a quality management system, doi:10.1016/j.scient.2011.05.021.

# CHAPTER 5

Ying Wang, Xiangyu Wang, Jun Wang, Ping Yung, and Guo Jun, Engagement of Facilities Management in Design Stage through BIM: Framework and a Case Study, doi.org/10.1155/2013/189105.

# CHAPTER 6

Yu-Cheng Lin and Yu-Chih Su, Developing Mobile- and BIM-Based Integrated Visual Facility Maintenance Management System, doi: 10.1155/2013/124249.

# CHAPTER 7

Shih-Ping Ho, Hui-Ping Tserng, and Shu-Hui Jan, Enhancing Knowledge Sharing Management Using BIM Technology in Construction, doi: 10.1155/2013/170498.

# CHAPTER 8

Yu-Cheng Lin, Yu-Chih Su, and Yen-Pei Chen, "Developing Mobile BIM/2D Barcode-Based Automated Facility Management System," The

Scientific World Journal, vol. 2014, Article ID 374735, 16 pages, 2014. doi:10.1155/2014/374735.

# CHAPTER 9

Gilberto A. Corona-Suárez, Simaan M. AbouRizk, and Stanislav Karapetrovic, Simulation-Based Fuzzy Logic Approach to Assessing the Effect of Project Quality Management on Construction Performance, doi.org/10.1155/2014/203427.

# CHAPTER 10

Mohammadreza Yadollahi, Mohammad Mirghasemi, Rosli Mohamad Zin, and Bachan Singh, Architect Critical Challenges as a Project Manager in Construction Projects: A Case Study, doi.org/10.1155/2014/205310

# Index

## C

## G



## H



## I



## J



## K



## L



## M



## N